D0072446

Mathematical Circles

Vol I

In Mathematical Circles
Quadrants I, II, III, IV

Mathematical Circles

Vol I

Howard Eves

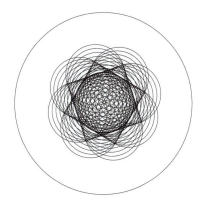

In Mathematical Circles
Quadrants I, II, III, IV

Published and Distributed by
The Mathematical Association of America

In Mathematical Circles: A Selection of Mathematical Stories and Anecdotes, Quadrants I and II and *In Mathematical Circles: A Selection of Mathematical Stories and Anecdotes, Quadrants III and IV* were previously published by Prindle, Weber & Schmidt, Incorporated in 1969.

© *2003 by*
The Mathematical Association of America, Inc.
Library of Congress Catalog Card Number 2002116306

ISBN 0-88385-542-9

Printed in the United States of America

Current Printing (last digit):
10 9 8 7 6 5 4 3 2 1

TO THE MATHEMATICS TEACHERS OF AMERICA
with so many of whom I have
had the pleasure of working

The Spectrum Series of the Mathematical Association of America was so named to reflect its purpose: to publish a broad range of books including biographies, accessible expositions of old or new mathematical ideas, reprints and revisions of excellent out-of-print books, popular works, and other monographs of high interest that will appeal to a broad range of readers, including students and teachers of mathematics, mathematical amateurs, and researchers.

777 Mathematical Conversation Starters, by John de Pillis

All the Math That's Fit to Print, by Keith Devlin

Circles: A Mathematical View, by Dan Pedoe

Complex Numbers and Geometry, by Liang-shin Hahn

Cryptology, by Albrecht Beutelspacher

Five Hundred Mathematical Challenges, Edward J. Barbeau, Murray S. Klamkin, and William O. J. Moser

From Zero to Infinity, by Constance Reid

The Golden Section, by Hans Walser. Translated from the original German by Peter Hilton, with the assistance of Jean Pedersen.

I Want to Be a Mathematician, by Paul R. Halmos

Journey into Geometries, by Marta Sved

JULIA: a life in mathematics, by Constance Reid

The Lighter Side of Mathematics: Proceedings of the Eugène Strens Memorial Conference on Recreational Mathematics & Its History, edited by Richard K. Guy and Robert E. Woodrow

MAA Service Center
P.O. Box 91112
Washington, DC 20090-1112
800-331-1622   FAX 301-206-9789

PUBLISHER'S NOTE

For many years Howard Eves, famed historian of mathematics and master teacher, collected stories and anecdotes about mathematics and mathematicians and gathered them together in six Mathematical Circles books. Thousands of teachers of mathematics have read these stories and anecdotes for their own enjoyment and used them in the classroom to add spice and entertainment, to introduce a human element, to inspire the student, and to forge some links of cultural history. Through a special arrangement with Professor Eves, the Mathematical Association of America (MAA) is proud to reissue all six of the Mathematical Circles books in this three-volume edition.

In Mathematical Circles, the first two books, were published to acclaim in 1969. They are bound together here as Volume I of the Mathematical Circles Collection. *Mathematical Circles Revisited* and *Mathematical Circles Squared* are bound together as Volume 2 of the Collection, and *Mathematical Circles Adieu* and *Return to Mathematical Circles* as Volume 3.

This three-volume set is a must for all who enjoy the mathematical enterprise, especially those who appreciate the human and cultural aspects of mathematics.

Ancient mazes consisted of a tortuous path confined to a small area of ground and leading to a tree or shrine in the center, with no chance of taking a wrong turn. Shown here is the circular maze constructed for the Minotaur. This labyrinth was delineated on the coins of Cnossus, specimens of which are not uncommon.

In Mathematical Circles

Quadrants I and II

Howard Eves

Published and Distributed by
The Mathematical Association of America

PREFACE

Somehow or other, over the years and without any particular effort on my part, a large number of stories and anecdotes about mathematics and mathematicians have fallen my way and remained stuck in my mind. These stories and anecdotes have proved very useful in the classroom-as little interest-rousing atoms, to add spice and a touch of entertainment, to introduce a human element, to inspire the student, to instill respect and admiration for the great creators, to yank back flagging interest, to forge some links of cultural history, or to underline some concept or idea. Many students and teachers have begged me to write up these stories and anecdotes, and a number of publishers have hounded me for them. At last I have given in and here offer a sample of the material.

Problems arose from the start. First of all, on marshalling the material I found I had far too much for a modest-sized venture; so I decided to select some three hundred to four hundred items as a test of reader interest. Next arose the problem of how to order the material; I decided to present it in rough chronological order with an accompanying index that would lend itself to other types of useful classification. And then there was the problem of the authenticity of some of the material; I decided not to make any effort at documentation, but simply to offer possibly doubtful items as part of the interesting accumulated folklore of our subject.

Undoubtedly many of the personal stories and anecdotes told here actually took place, but it is equally certain that some originally true stories have been embroidered over the years and ages, and that others have simply been made up as being apposite to the subjects concerned. Thus there are stories that have come down to us about some great men so lost in the mythical haze of the past that really nothing certain can be told about them; there are identical anecdotes that have been told about different persons; there are amusing tales that have circulated but have been denied by the principals involved; there are many cases where the same basic story has been told in varying and sometimes conflicting versions. One is reminded of Abraham Lincoln. There are literally hundreds of anecdotes that have been told about Lincoln; many of these have a real basis, but there can be no doubt that some of them were embroidered, twisted, or simply devised to fit the interesting and colorful character of Lincoln.

Particularly difficult is the matter of anecdotes about contemporary people. My collection contains a large number of such stories, but I have heard some of them denied or at least made much less interesting by the principals themselves. So, in this first round of stories, I shall stick to the past wherein the *dramatis personae* cannot rise up and defend themselves, and I shall refrain from narrating any anecdotes about the living.

The bulk of the material can be read with very little, and most often with no, mathematical background. But here and there do occur items a bit more demanding, and there are even a few challenging elementary problems. The historical comments and capsules are largely adapted from my book, *An Introduction to the History of Mathematics* (Holt, Rinehart and Winston, third edition, 1969). There the interested reader can find fuller historical treatments. I am very grateful to *The Mathematics Teacher*, one of the fine official journals of the National Council of Teachers of Mathematics, for permitting me to reproduce in essentially the original form certain items which appeared there in the Historically Speaking section, a department of the journal that for some years I have had the pleasure of editing.

It is hoped that the general reader may find the potpourri sufficiently savory, that the teacher may find it useful to serve on occasion, and that the partaking student may enjoy some of the historical tidbits and (in a non-cannibalistic way) the human flavor.

With sufficient encouragement, I may decide to travel around the Mathematical Circle, or at least the more modern part of it, again in the future. Toward this possibility, interested readers are cordially invited to contribute any favorite stories they would like to see in an expanded collection.

HOWARD W. EVES

CONTENTS

CONTENTS

CONTENTS

CONTENTS

QUADRANT TWO

Contents

Contents

QUADRANT ONE

From a mathematical crow
to Hypatia's tragic death

THE ANIMAL WORLD, REAL AND IMAGINARY

FAR, far back in time, before the appearance of Homo sapiens on the earth, there already were animals of various sizes and habits. Did any of these creatures possess even a tinge of mathematical sense? From the study, by competent observers, of present-day animal behavior, there has accumulated a weighty mass of evidence supporting the belief that certain birds and certain arthropods perhaps have such a quality. The conclusions are controversial and other plausible explanations have been offered. But many spiders do make beautifully symmetric webs of almost-perfect regular polygons, honey bees have long attracted admiration for the construction of their hexagonal comb-cells, some insects seem to possess an uncanny number sense when laying eggs, and most birds realize something is different when a sufficient number of eggs are added to or taken from their nests.

1° *A Scotch crow.* There is a touching and authentic story about a bird that seemed to possess a number sense. A squire in Scotland became annoyed by a raucous crow that had made its nest in the watchtower of his estate, and he determined to shoot the bird. Repeatedly he tried to enter the tower to blast the bird, but each time at the man's approach the crow would leave its nest and take up a watchful position in a distant tree. When the wearied squire would finally leave the tower, the bird would return to its nest. Not wishing to be outsmarted by a bird-brain, the squire resorted to a ruse. He secured the assistance of a neighbor one day. The two men entered the tower, one man came out and went away, and the other remained within. But the crow was not deceived; it stayed in the distant tree until the man within the tower came out. The experiment now became a contest, and the next day three men entered the tower, two came out and went away, and the third waited within to blast the pesky bird. But the crow was not fooled; it remained in the distant tree until the man within the tower came out. The next day the experiment was repeated with four men, but still without success. Finally five men entered the tower, four came out and went away, and the fifth remained inside. At this point the crow seemed to have lost count and,

3

unable to distinguish between four and five, it returned to its nest in the tower.

The conclusion of the story has not been recorded, but it is hoped that by the time of the final experiment a sufficient affection and respect had been built up for the crow so that the bird was allowed to remain nesting in the tower.

2° *The solitary wasp.* A striking instance of what may be number sense in insects is illustrated by the so-called *solitary wasp*. The mother wasp lays her eggs individually in separate cells and then provides each cell with a number of live caterpillars on which the young feed when they hatch. The remarkable thing is that the number of caterpillars is surprisingly uniform for a given species of wasp—some species provide five per cell, others twelve, and still others as many as twenty-four. Most surprising is the genus *eumenus*, a variety in which the female is much larger than the male. Somehow or other, the mother wasp knows whether the egg will produce a female or a male grub; if the egg is female she provides its cell with ten caterpillars, if the egg is male she provides its cell with five.

3° *The Harvard katydid.* The physicist, Professor George W. Pierce, of the Cruft Laboratory at Harvard University, has studied the songs of insects. There was a katydid in Dr. Pierce's laboratory that learned to count and thereby alter its usual two-beat rhythm. During an experiment, a laboratory assistant who could imitate the katydid's shrill "zeep-zeep," made the sound in three beats instead of two. The katydid answered with three beats. The assistant then tried four, and the katydid answered with four. Then the assistant tried five and the katydid answered with five. At the next stage, however, the insect lost count and, on its own, began to improvise on the numbers it had already learned.

4° *The ingenious honey bees.* Man has long shown interest in the seeming geometrical sagacity of the honey bees. The first known man to report on this mathematical acumen was the eminent Greek geometer Pappus, who flourished in Alexandria some sixteen hundred years ago. In Book V of his famous *Mathematical Collection* we find the following passage concerning the extremum properties of the cells of the bees' honeycombs:

> Presumably because they know themselves to be entrusted with the task of bringing from the gods to the accomplished portion of mankind a share of ambrosia in this form, they do not think it proper to pour it carelessly on ground or wood, or any other ugly or irregular material; but first collecting the sweets of the most beautiful flowers which grow upon the earth, they make from them for the reception of the honey, the vessels which we call honeycombs (with cells) all equal and similar, and contiguous to one another, and hexagonal in form. And that they have contrived this by virtue of a certain geometrical forethought, we may infer in this way. They would necessarily think that the figures must be such as to be contiguous to one another, that is to say, to have their sides in common in order that no foreign matter could enter into the interstices between them and so defile the purity of the produce. Now there are three rectilinear figures which are capable of fulfilling this condition, I mean regular figures which are equilateral and equiangular, for bees would have none of figures which are not uniform. . . . There being then three figures capable by themselves of exactly filling up the space about the same point, the bees by reason of their instinctive wisdom chose for their construction the figure which has most angles, because they conceived it would hold more honey than either of the other two.

5° *A classification of mathematicians.* Francis Bacon (1561–1626), the English moralist, prophet, philosopher, and man of letters, often engaged in scientific writings studded with aphorisms, many of which are particularly applicable to mathematics and mathematicians. For example, he divided philosophers into three groups—the ants, the spiders, and the bees. The ants are those who diligently but stupidly and unsystematically gather many little and generally useless bits of knowledge; the spiders are those who spin out intricate and insubstantial theories from their own minds; the bees are those who go to nature for raw material and inspiration, and through exacting labor transfer these into sound theories. These last he called the true philosophers. One can pretty well here replace "philosophers" with "mathematicians."

6° *Logarithms and multiplication.* When teaching logarithms in a trigonometry or algebra course, one can make a point and at the same time entertain the class with the following whimsical story.

One of our great national parks was yearly visited by a certain man. On one of these visits, the man met a snake and the snake's wife, but saw no little snakes. Accordingly, in conversation with the snakes, the man asked, "How come there are no little snakes?" "Well, you see," replied Mr. Snake, "we are adders, and cannot multiply." The following year, upon returning to the park, the man again found Mr. and Mrs. Snake, but now there were many little snakes. "How come there are so many little snakes?" the man asked. "Well, you see," replied Mr. Snake, "the park ranger came through here and built us a log table, so now we adders can multiply."

7° *Good induction versus bad induction.* A scientist had two large jars before him on the laboratory table. The jar on his left contained a hundred fleas; the jar on his right was empty. The scientist carefully lifted a flea from the jar on the left, placed the flea on the table between the two jars, stepped back and in a loud voice said, "Jump." The flea jumped and was put in the jar on the right. A second flea was carefully lifted from the jar on the left and placed on the table between the two jars. Again the scientist stepped back and in a loud voice said, "Jump." The flea jumped and was put in the jar on the right. In the same manner, the scientist treated each of the hundred fleas in the jar on the left, and each flea jumped as ordered. The two jars were then interchanged and the experiment continued with a slight difference. This time the scientist carefully lifted a flea from the jar on the left, *yanked off its hind legs*, placed the flea on the table between the jars, stepped back and in a loud voice said, "Jump." The flea did not jump, and was put in the jar on the right. A second flea was carefully lifted from the jar on the left, its hind legs yanked off, and then placed on the table between the two jars. Again the scientist stepped back and in a loud voice said, "Jump." The flea did not jump, and was put in the jar on the right. In this manner, the scientist treated each of the hundred fleas in the jar on the left, and in no case did a flea jump when ordered. So the scientist recorded in his notebook the following induction: "A flea, if its hind legs are yanked off, cannot hear."

8° *The mathematical horse.* There was a horse that showed a remarkable ability at learning mathematics. The horse mastered arithmetic, and then elementary algebra. Soon after, it acquired plane and solid geometry, and next, trigonometry. Then it was offered analytic geometry, but upon this the horse balked, kicked, and carried on in a wild manner. All of which simply proves that one mustn't put Descartes before the horse.

PRIMITIVE MAN

THERE are stories gathered by explorers and anthropologists pointing up the meagerness of mathematical sense among very primitive peoples. It seems that the first mathematical considerations of man probably had their origin in simple observations stemming from human ability to recognize physical form and to compare shapes and sizes. This very early mathematics may be called *subconscious mathematics*.

9° *Two plus two.* Sir Francis Galton (1822–1911), English scientist, explorer, and anthropometrist, has related that the primitive Damaras of Africa, in bartering two sticks of tobacco for one sheep as the rate of exchange, became hopelessly confused when a white trader, desiring two sheep, offered four sticks of tobacco at once. Fraud was suspected by the Damaras, and the transaction had to be revised and carried out more slowly. First two sticks of tobacco were given and one sheep driven away, then two more sticks of tobacco and the second sheep claimed. When shown that the result came out the same as the trader's original proposal, the tribesmen regarded the trader as one possessed of magic powers.

Yet, these Damaras were not unintelligent. They knew precisely the size of a flock of sheep or a herd of oxen, and would miss an individual at once, because they knew the faces of all of the animals. To us, this form of intelligence, which is true and keen observation, would be infinitely more difficult to cultivate than that involved in counting.

10° *Addition of vectors.* Two vectors cannot be added by simply adding their magnitudes, inasmuch as the directions of the vectors are important in this consideration. An incredible story concerning the addition of vectors has been told by Martin Johnson, the famous world traveler. One day, in central Africa, Johnson came upon eight husky tribesmen on the verge of exhaustion as four pushed a mired Land Rover from the front and four pushed it from the rear without moving the vehicle an inch. The natives were much impressed with Johnson's ability to rearrange vectors when he placed all eight men behind the Land Rover and it moved forward quite easily.

11° *The great size of three.*   In the development, among primitive people, of vocal sounds to represent successively the first few small numbers, a sound is often finally reached which merely signifies "a great many." This, for example, is the case of those natives of Queensland who count "one, two, two and one, two twos, much," where the last term is meant to cover not only five, but all numbers larger than four. It is interesting that there are instances of the use of the very small number three in this sense of "much," or an excessively large number. For instance, the native Tasmanians count "one, two, plenty." Again, it is this use of three that we find in the Latin phrase "ter felix," which, though it literally means "thrice happy," is really meant to imply "very happy." It occurs again in the English "Thrice is he armed that hath his quarrel just," and in the French "très bien."

12° *Gog and Gug.*   There are a number of fanciful mathematical stories, of varying quality, about prehistoric or primitive people. Two of these people go by the names of Gog and Gug, and the following (which has appeared in several versions) is representative of the Gog and Gug stories.

In a certain tribe, in which polygamy was practiced, a married man's standing in the tribe depended upon the combined weight of his wives—the greater the combined weight, the more important was the man. Every year, on weighing day and according to custom, the married men would stand their wives on neatly spread animal skins, and the chief of the tribe would come around with a crude seesaw and balance the wives of one man against those of another in order to determine the relative importance of the men. Now Gog had only one wife, who was very heavy, while Gug had two much slenderer wives, and all year the two men argued as to who was the more important. When weighing day arrived, Gog placed his wife on a large hippopotamus skin, and Gug placed his wives on two small gazelle skins. When the weighing was performed, it was found that Gog's wife exactly balanced against the two wives of Gug. Thus it turned out that the two men were equally important, since, by the chief's ruling, "the squaw on the hippopotamus is equal to the sum of the squaws on the other two hides."

PRE-HELLENIC MATHEMATICS

IN the beginning, man considered only concrete mathematical problems, which presented themselves individually and with no observed interconnections. When human intelligence was able to extract from a concrete mathematical relationship a general abstract relationship containing the former as a particular case, mathematics became a science. This stage of mathematics may be called *empirical*, or *scientific*, *mathematics*, for mathematical findings were discovered by trial and error, experimentation, and other empirical or laboratory-type procedures.

There is no evidence that permits us to estimate the number of centuries that passed before man was able to raise mathematics from its subconscious stage to the scientific stage. As far back as history allows us to grope into the past, we find already present a sizeable body of scientific mathematics. There is historical evidence that this type of mathematics arose in certain river basins of the ancient Orient that cradled advanced forms of society. Among these river basins were the Nile in Egypt, the Tigris and Euphrates of Mesopotamia, the Indus and Ganges of south-central Asia, and the Hwang Ho and the Yangtze of eastern Asia. The deductive element is almost completely lacking in this early mathematics. It was not until the days of the Greeks, starting about 600 B.C., that deduction was made to play an essential role in mathematics, and mathematics became *systematic*, or *deductive*, *mathematics*.

The name of no mathematician of the pre-Greek period has come down to us, and so, of course, no personal mathematical anecdotes can be told for this era. Nevertheless, there are a number of interesting stories related to the mathematics of the period. Indeed, it would be difficult to find a more exciting mathematical story than that of Plimpton 322 (see Item 17°).

13° *Problem 79 of the Rhind papyrus.*   One of our chief primary sources concerning the mathematics of ancient Egypt is the Rhind, or Ahmes, papyrus, a mathematical text partaking of the nature of a practical handbook and consisting of eighty-five problems copied about 1650 B.C. in hieratic writing by the scribe Ahmes from an earlier work. The papyrus was purchased in Egypt by the English Egyptologist A. Henry Rhind and then later acquired by the British Museum.

Although little difficulty was encountered in deciphering and then interpreting most of the problems in the Rhind papyrus, there is one problem, Problem Number 79, for which the interpretation is not so certain. In this problem occurs the following curious set of data, here transcribed:

	Estate
Houses	7
Cats	49
Mice	343
Heads of wheat	2401
Hekat measures	16807
	19607

One easily recognizes the numbers as the first five powers of seven, along with their sum. Because of this it was at first thought that perhaps the writer was here introducing the symbolic terminology *houses, cats,* and so on, for *first power, second power,* and so on.

A more plausible and interesting explanation, however, was given by the historian Moritz Cantor in 1907. He saw in this problem an ancient forerunner of a problem that was popular in the Middle Ages, and which was given by Leonardo Fibonacci in 1202 in his *Liber abaci.* Among the many problems occurring in this work is the following: "There are seven old women on the road to Rome. Each woman has seven mules; each mule carries seven sacks; each sack contains seven loaves; with each loaf are seven knives; and each knife is in seven sheaths. Women, mules, sacks, loaves, knives, and sheaths, how many are there in all on the road to Rome?'' As a later and more familiar version of the same problem we have the old English children's rhyme:

> As I was going to St. Ives
> I met a man with seven wives;
> Every wife had seven sacks;
> Every sack had seven cats;
> Every cat had seven kits.
> Kits, cats, sacks, and wives,
> How many were going to St. Ives?

According to Cantor's interpretation, the original problem in the Rhind papyrus might then be formulated somewhat as follows: "An estate consisted of seven houses; each house had seven cats; each cat ate seven mice; each mouse ate seven heads of wheat; and each head of wheat was capable of yielding seven hekat measures of grain. Houses, cats, mice, heads of wheat, and hekat measures of grain, how many of these in all were in the estate?"

Here, then, may be a problem that has been preserved as part of the puzzle lore of the world. It was apparently already old when Ahmes copied it, and older by close to three thousand years when Fibonacci incorporated a version of it in his *Liber abaci*. More than seven hundred and fifty years later we are reading another variant of it to our children. One cannot help wondering if a surprise twist such as occurs in the old English rhyme may also have occurred in the ancient Egyptian problem, though, in all likelihood, this twist was an Anglo-Saxon contribution.

There are many puzzle problems popping up every now and then in our present-day magazines that have medieval counterparts. How much further back some of them go is now almost impossible to determine.

14° *The pyramid of Gizeh.* The great pyramid of Gizeh was erected about 2900 B.C. and undoubtedly involved some mathematical and engineering problems. The structure covers thirteen acres and contains over two million stone blocks, averaging two and a half tons in weight, very carefully fitted together. These stone blocks were brought from sandstone quarries located on the other side of the Nile. Some chamber roofs are made of fifty-four-ton granite blocks, twenty-seven feet long and four feet thick, hauled from a quarry six hundred miles away, and set two hundred feet above ground. It is reported that the sides of the square base of the pyramid involve a relative error of less than one fourteen thousandth, and that the relative error in the right angles at the corners does not exceed one twenty-seven thousandth.

Of course, the engineering skill implied by these impressive statistics is considerably diminished when we realize that the task was accomplished by an army of 100,000 laborers working for a period of thirty years. For example, the seemingly difficult problem of raising the great stone blocks to ever higher positions as the structure of the pyramid grew was probably very simply met by constantly keeping the rising pyramid submerged in imported sand, hauling the blocks on rollers up a gradual incline of this sand, and then finally removing all the imported sand. Given sufficient time and sufficient labor power, many difficult feats can be done in a simple and primitive fashion. Nevertheless, the engineering problems had to be overcome one way or another. Consider, for instance, the mammoth engineering problems encountered by the Egyptians in quarrying and setting up some of their huge obelisks of pink granite. The largest existing obelisk appears before the old Temple of the Sun at Thebes; it was quarried about 1500 B.C. and is no less than 105 feet long, nearly 10 feet square at the base, and weighs about 430 tons!

15° *The greatest Egyptian pyramid.*   In the Moscow papyrus, an ancient Egyptian text dating from about 1850 B.C. and consisting of twenty-five mathematical problems, we find the following numerical example:

> If you are told: A truncated pyramid of 6 for the vertical height by 4 on the base by 2 on the top. You are to square this 4, result 16. You are to double 4, result 8. You are to square 2, result 4. You are to add the 16, the 8, and the 4, result 28. You are to take one third of 6, result 2. You are to take 28 twice, result 56. See, it is 56. You will find it right.

Now if we recall that the ancient Egyptians concerned themselves with only *square* pyramids, this certainly seems to be a numerical example illustrating the use of the formula

$$V = \tfrac{1}{3}h(a^2 + ab + b^2)$$

for computing the volume of a frustum of a square pyramid of height h and with bases of sides a and b. In the example, $h = 6$, $a = 4$, $b = 2$. The instructions of the problem carry us step by step through the substitution of these values for a, b, h in the above formula, resulting in the correct answer of 56 for the volume of the frustum. If this interpretation is the right one (and it is difficult to imagine that it is not), then we must concede to the ancient Egyptians the extraordinary achievement of the discovery of the formula. Now any rigorous derivation of the formula requires some form of the integral calculus, and so the discovery must have been an inductive, or empirical, one. No other unquestionably genuine example of this formula has been found in pre-Hellenic mathematics, and several involved conjectures have been furnished to explain how it might have been empirically discovered. It appears to be such a magnificent piece of induction that the historian of mathematics E. T. Bell has aptly referred to this early Egyptian example as "the greatest Egyptian pyramid," since the inductive discovery involved is perhaps more remarkable than the physical construction of any of the great stone pyramids of antiquity.

16° *Squaring the circle.*   The earliest attempted solution that has come down to us of the problem of constructing a square equal in area to a given circle is found in the Rhind papyrus of about 1650 B.C. There we are told that the area of a circle is equal to that of the square having $\tfrac{8}{9}$ths of the diameter of the circle as a side. It is an elementary matter to show that this solution of the problem is equivalent to taking $\pi = (\tfrac{4}{3})^4$ or, approximately, 3.16.

One naturally wonders how the Egyptians arrived at the above approximate solution of the circle-squaring problem. In all likelihood, it was obtained by some sort of empirical procedure similar to the following: Draw a large circle on flat ground and cover its interior tightly with a large number of small pebbles, and only one pebble high, chosen as uniform in shape and size as possible. After this is done, convert the collection of pebbles from their circular array into a square array. On measuring the side of the square one will find it to be very close to $\frac{8}{9}$ths of the diameter of the original circle.

17° *Plimpton 322.* Archeologists working in Mesopotamia have systematically unearthed, since before the middle of the nineteenth century, some half-million inscribed clay tablets. Of these half-million tablets, about three hundred have so far been identified as strictly mathematical tablets containing mathematical tables and lists of mathematical problems. We owe most of our knowledge of ancient Babylonian mathematics to the scholarly deciphering and interpretation of many of these mathematical tablets.

Perhaps the most remarkable of the Babylonian mathematical tablets yet analyzed is that known as Plimpton 322, meaning that it is the item with catalogue number 322 in the G. A. Plimpton collection at Columbia University. The tablet is written in Old Babylonian script, which dates it somewhere from 1900 to 1600 B.C., and it was first described by Otto Neugebauer and A. J. Sachs in 1945.

Figure 1 gives an idea of the shape of the tablet. Unfortunately a

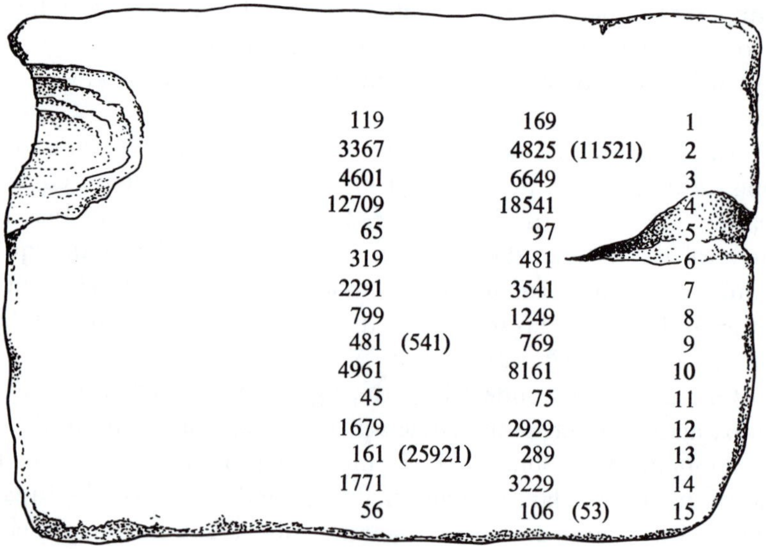

119		169		1
3367		4825	(11521)	2
4601		6649		3
12709		18541		4
65		97		5
319		481		6
2291		3541		7
799		1249		8
481	(541)	769		9
4961		8161		10
45		75		11
1679		2929		12
161	(25921)	289		13
1771		3229		14
56		106	(53)	15

FIGURE 1

missing piece has been broken from the entire left edge and the tablet
is further marred by a deep chip near the middle of the right edge and
a flaked area in the top left corner. Upon examination, crystals of
modern glue were found along the left broken edge of the tablet. This
suggests that the tablet was probably complete when excavated, that
it subsequently broke, that an attempt was made to glue the pieces
back together, and that later the pieces again separated. Thus the
missing piece of the tablet may still be in existence but, like a needle in
a haystack, lost somewhere among the collections of these ancient
tablets. As we shall shortly see, it would be very interesting if this
missing piece should be found.

The tablet contains three essentially complete columns of figures
which, for convenience, are reproduced on Figure 1 in our own decimal
notation. There is a fourth and partly incomplete column of figures
along the broken edge. We shall later reconstruct this column.

It is clear that the column on the extreme right merely serves to
number the lines. The next two columns seem, at first glance, to be
rather haphazard. With study, however, one discovers that corresponding numbers in these columns, with four unfortunate exceptions,
constitute the hypotenuse and a leg of integral-sided right triangles.
The four exceptions are noted in Figure 1 by placing the original
readings in parentheses to the right of the corrected readings. The
exception in the second line has received an involved explanation, but
the other three exceptions can be easily accounted for. Thus, in the
ninth line, 481 and 541 appear as (8, 1) and (9, 1) in the sexagesimal
system. Clearly the occurrence of 9 instead of 8 could be a mere slip
of the stylus when writing these numbers in cuneiform script. The
number in line 13 is the square of the corrected value, and that in the
last line is half of the corrected value, showing that the squares and
the halves of certain numbers in the table probably played a role in
the construction of the table.

Now a set of three positive integers, like (3, 4, 5), which can be the
sides of a right triangle, is known as a *Pythagorean triple*. Again, if the
triple contains no common factor other than unity, it is known as a
primitive Pythagorean triple. Thus (3, 4, 5) is a primitive triple, whereas
(6, 8, 10) is not. One of the achievements of the Arabians, two thousand
years after the date of the Plimpton tablet, was to show that all primitive

Pythagorean triples (a, b, c) are given parametrically by

$$a = 2uv, \qquad b = u^2 - v^2, \qquad c = u^2 + v^2,$$

where u and v are relatively prime, of different parity, and $u > v$. Thus if $u = 2$ and $v = 1$, we obtain the primitive triple $a = 4$, $b = 3$, $c = 5$.

Suppose we compute the other leg a of the integral-sided right triangles determined by the given hypotenuse c and leg b on the Plimpton tablet. We find the following Pythagorean triples:

lines	a	b	c	u	v
1	120	119	169	12	5
2	3456	3367	4825	64	27
3	4800	4601	6649	75	32
4	13500	12709	18541	125	54
5	72	65	97	9	4
6	360	319	481	20	9
7	2700	2291	3541	54	25
8	960	799	1249	32	15
9	600	481	769	25	12
10	6480	4961	8161	81	40
11	60	45	75	2	1
12	2400	1679	2929	48	25
13	240	161	289	15	8
14	2700	1771	3229	50	27
15	90	56	106	9	5

It will be noticed that all of these triples, except the ones in lines 11 and 15, are primitive triples. For discussion we have also listed the values of the parameters u and v leading to these Pythagorean triples. The evidence seems good that the Babylonians of this remote period were acquainted with the general parametric representation of primitive Pythagorean triples as given above. This evidence is strengthened when we notice that u and v, and hence also a (since $a = 2uv$), are *regular* sexagesimal numbers—that is, are numbers of the form $2^p 3^q 5^r$ and thus have their reciprocals expressible as terminating sexagesimal fractions. It appears that the table on the tablet was constructed by deliberately choosing small regular numbers for the parameters u and v.

This choice of u and v must have been motivated by some subsequent process involving division, for regular numbers appear in tables of reciprocals and are useful in reducing division to multiplication. An examination of the fourth, and partially destroyed, column gives the answer. For this column is found to contain the values of $(c/a)^2$ for the different triangles. To carry out the division, the side a, and hence the numbers u and v, had to be regular.

It is worth examining the column of values for $(c/a)^2$ a little more deeply. This column, of course, is a table giving the square of the secant of the angle B opposite side b of the right triangle. Because side a is regular, sec B has a finite sexagesimal expansion. Moreover it turns out, with the particular choice of triangles as given, that the values of sec B form a surprisingly regular sequence which decreases by almost exactly $\frac{1}{60}$ as we pass from one line of the table to the next, and the corresponding angle decreases from 45° to 31°. We thus have a secant table for angles from 45° to 31°, formed by means of integral-sided right triangles, in which there is a uniform jump in the function rather than in the corresponding angle. All this is truly remarkable. It seems highly probable that there were companion tables giving similar information for angles ranging from 30° to 16° and from 15° to 1°.

The analysis of Plimpton 322 shows the careful examination to which some of the Babylonian mathematical tablets must be subjected. Formerly such a tablet might have been summarily dismissed as merely a business list or record.

18° *The angular degree.* We undoubtedly owe to the ancient Babylonians our present division of the circumference of a circle into 360 equal parts. Several explanations have been put forward to account for the choice of this number.

Thus it has been supposed that the early Babylonians reckoned a year as 360 days, and that this naturally led to the division of the circle into 360 equal parts, each part representing the daily amount of the supposed yearly revolution of the sun about the earth. Today this explanation is discredited, since we now have evidence that the Babylonians very early knew that the year possesses more than 360 days.

Again, the early Babylonians were very likely familiar with the geometric fact that the radius of a circle can be applied exactly six times to its circumference as a chord. Then, having adopted sexagesimal division, it was natural to divide the central angles of these chords into 60 equal parts, resulting in the division of the entire circle into 360 equal parts.

Otto Neugebauer, the great scholar and authority on early Babylonian mathematics and astronomy, has proposed an interesting alternative explanation. In early Sumerian times there existed a large distance unit, a sort of *Babylonian mile*, equal to about seven of our miles. Since the Babylonian mile was used for measuring longer distances, it was natural that it should also become a time unit, namely the time required to travel a Babylonian mile. Later, sometime in the first millennium B.C., when Babylonian astronomy reached the stage in which systematic records of celestial phenomena were kept, the Babylonian time-mile was adopted for measuring spans of time. Since a complete day was found to be equal to 12 time-miles, and one complete day is equivalent to one revolution of the sky, a complete circuit was divided into 12 equal parts. But, for convenience, the Babylonian mile had been subdivided into 30 equal parts. We thus arrive at $(12)(30) = 360$ equal parts in a complete circuit.

19° *Magic squares.* There are some Chinese mathematical works of which parts, at least, are claimed to date from very early times. This is difficult to verify because we lack original sources. As an added complication it was decreed, in 213 B.C. by the Emperor Shi Huang-ti, that all books in the country be burned and all scholars be buried. Although the edict was most certainly not completely carried out, and many books that were burned were soon restored from memory, we are now in doubt as to the genuineness of anything claimed to be older than the unfortunate date.

One of the oldest of the Chinese mathematical classics is the *I-king*, or *Book on Permutations*. In this appears a numbered diagram, known as the *lo-shu*, later pictured as in Figure 2.

The lo-shu is the oldest known example of a magic square, and myth claims that it was first seen by the Emperor Yu, in about 2200 B.C., as a decoration upon the back of a divine tortoise along a bank of the Yellow River. It is a square array of numerals indicated in Figure 2 by knots in strings, black knots for even numbers and white

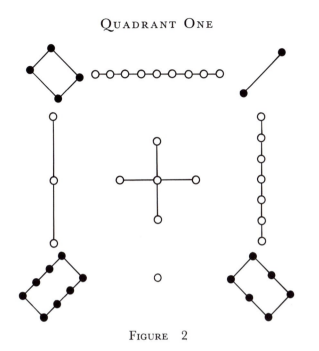

FIGURE   2

knots for odd numbers. The sum of the three numbers in any column is equal to the sum of the three numbers in any row, and also to the sum of the three numbers in either diagonal.

20° *The 3–4–5 triangle problem.* There are reports that ancient Egyptian surveyors laid out right angles by constructing 3–4–5 triangles with a rope divided into 12 equal parts by 11 knots. Since there is no documentary evidence to the effect that the Egyptians were aware of even a particular case of the Pythagorean theorem, the following purely academic problem arises: Show, without using the Pythagorean theorem, its converse, or any of its consequences, that the 3–4–5 triangle is a right triangle. The reader may care to try to solve this problem. Many different solutions have been given over the years. A particularly simple solution can be effected based on Figure 3, which shows four right triangles with legs of 3 and 4 units, along with a unit square, making up a square of 25 square units of area. This means that the hypotenuse of a right triangle with legs of 3 and 4 units is 5 units long, and it follows that a 3–4–5 triangle is a right triangle. This solution is found in the *Chou-pei Suan-king* (*The Arithmetical Classic of the*

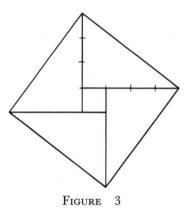

FIGURE 3

Gnomon and the Circular Paths of Heaven), a work generally regarded as the oldest of the ancient Chinese mathematical classics. The solution is easily generalized to obtain the Pythagorean relation for an arbitrary right triangle.

A FEW LATER CHINESE STORIES

THOUGH the following three stories of long-ago Chinese mathematicians postdate the Greek period, it seems convenient to place them here. The first story is a brief and sad biography; the other two are charming and semifanciful tales.

21° *The Drunken Dragon loses his hair.* The mathematician Ts'ai Yung, one of many Chinese calendric experts, flourished about 190 A.D. All of his works are lost to us. It has been reported that his persistent habit of conviviality earned him the name Drunken Dragon, and that at one time he was sentenced to death for political reasons, but at the last moment his sentence was commuted to having his hair pulled out.

22° *Huai-Wen calculates the dates on a tree.* Some of the early Chinese mathematicians became quite adept with the abacus and the calculating rods. There is a story that at the college at Chinyang, sometime around the year 560, a foreign Buddhist monk exclaimed over the calculating ability of the mathematician and metallurgist Huai-Wen.

Pointing to a jujube tree in the courtyard of the college, the monk asked Huai-Wen to calculate the number of dates on the tree. With the aid of his calculating device, Huai-Wen soon stated not only the total number of dates on the tree, but also how many were ripe, how many unripe, and how many only partly ripe. When, to test Huai-Wen's results, the dates were collected and counted, it was found that the mathematician was one date short. "It cannot be," said Huai-Wen, "shake the tree once more." And sure enough, one more date fell to the ground.

23° *I-Hsing finds his teacher.* The most famous of the Thang mathematicians was the monk I-Hsing, who flourished about 725 A.D., and who, by imperial order, once prepared a calendar. All of his books are lost. In the *Ming Huang Tsa Lu* of Cheng Chhu-Hui, written in 855, there is a brief account of I-Hsing's life.

It seems that before I-Hsing was introduced to the emperor, he had studied under Phu-Chu at Sung Shan. During an entertainment of monks there, a very learned member of the party, named Lu Hung, wrote an essay commemorating the meeting. In writing the essay, Lu Hung used very difficult words, and he announced that he would take as his pupil any student present who could read and understand the essay. I-Hsing stepped forward, glanced quickly through the essay, and then smilingly laid it down. Lu Hung was annoyed by I-Hsing's offhand manner, but when I-Hsing repeated the essay without a single mistake, Lu Hung was overcome and told Phu-Chu that this student was not one to be taught, but that he had better be allowed to travel.

So I-Hsing, wishing to study indeterminate analysis, traveled far and wide seeking an appropriate instructor. In time he came to the remote astronomical observatory at the Kuo Chhing Ssu temple, before which there was a courtyard with a spring flowing in it. As I-Hsing stood in the courtyard, he overheard an old monk inside the temple say, "Today someone will arrive to learn my mathematical art. Indeed, he should be at the door by now. Why doesn't someone bring him in?" Shortly the monk spoke aloud again, saying, "In the courtyard the waters of the spring are flowing westward—my student should be arriving." So I-Hsing entered the temple and knelt before the monk, who then and there began to teach the student his computing methods, whereupon the waters of the spring in the courtyard immediately turned and flowed eastward.

This story points up the difficulties of mathematical communication in those early days, and it shows how easily mathematical discoveries might die with the author.

THALES

ACCORDING to tradition, Greek mathematics appears to have started in an essential way with the work of Thales of Miletus in the first half of the sixth century B.C. This versatile genius, declared to be one of the "seven wise men" of antiquity, was a worthy founder of *systematic mathematics* and is the first known individual with whom the use of deductive methods in mathematics is associated. As with other great men, many charming anecdotes are told about Thales, which, if not true, are at least apposite.

24° *How to become rich.* One day, when still a poor man, Thales was conversing with an equally poor visiting friend. "The lot of a poor man is a hard one in this world," remarked the friend, "and if one is born poor one will remain poor the rest of his life." "Not necessarily so," replied Thales, "I am sure that if one should apply himself to the matter, it would be easy to become rich." "That is certainly easier said than done," rejoined the friend, "and I find it very difficult to believe." "I tell you what," said Thales, "when you visit me again six months from now, I will show you how easy it is to become rich." Six months later, when the visitor returned, he was astonished to find that his former poor friend was now the richest man for miles around, and he expressed his great surprise to Thales. "I merely wanted to show you how easy it is to become rich if you apply yourself to the problem," replied Thales. "Well, tell me, how did you do it?" asked the friend. "It was simple," explained Thales. "Foreseeing a heavy crop of olives coming, I went about and secretly bought up all the olive presses of the region. When it came time to press the oil from the olives, no one in the region had a press, and all had to come to me. I realized a fortune by renting out the presses. You see, it is easy to become rich if you but give the matter a little attention."

25° *The recalcitrant mule.* As one of his smaller business ventures, Thales had a salt mine up in the hills. Each day his mules came down the hill carrying bags of salt from the mine. The trail down the hill crossed a small stream, and one day while fording the stream one of the laden mules slipped and fell. In struggling to its feet, the mule chanced to roll onto its back and the salt dissolved from its bags, leaving the mule light and comfortable for the rest of the trip down the hill. The mule remembered this experience, and on the three successive days when it came down the hill it rolled about in the stream to dissolve and lose the contents of its load. Now ordinary mortals, like you and me, would have seized a stick and, with much shouting, would have beaten the animal over its hind quarters to cure it of this troublesome habit. But Thales was a "wise man," and the next time that mule came down the hill it was loaded with bags of sponges.

26° *Why Thales never married.* In his later and more affluent period of life, Thales was visited by his friend Solon, the great Greek lawgiver. As the two men sat discoursing one evening, Solon queried, "Here you are in possession of almost everything a man might desire— wealth and influence, health and comfort, fame and respect, knowledge and wisdom—but no wife. Tell me, my friend, why have you never married?" "That is quite a question," replied Thales, "and since the hour is so late, I believe I can better tell you in the morning." So the host and his guest retired for the night. In the morning they assembled for breakfast, but they had scarcely begun to eat their bowls of grapes when in raced a breathless runner with a message for Solon. Solon read the message and rose grief-stricken to his feet. "Thales," he hoarsely said, "I must leave at once for home. This message informs me that my favorite son has fallen from his horse and been killed." "There, there, my dear Solon," replied Thales soothingly, "please be seated and calm yourself. The message is a fictitious one, and I have contrived this whole event. I merely wanted to tell you why I never married."

27° *Thales as a stargazer.* One evening, while absently walking along and studying the stars, Thales inadvertently fell into a deep ditch from which he could not extricate himself. Finally his calls for help were answered by an old woman who managed to pull him from the ditch. When asked by the old woman what he had been doing to get himself into such a predicament, he explained that he had been looking at the stars, whereupon the old woman wondered, "How can you hope to see anything in the heavens when you can't even see what is at your own feet?"

28° *Credit where credit is due.*   When once asked what he would take for one of his discoveries, Thales replied, "I will be sufficiently rewarded if, when telling it to others, you will not claim the discovery as your own, but will say it was mine."

29° *Moral advice.*   When Thales was asked how we might lead more upright lives, he advised, "By refraining from doing what we blame in others."

30° *An incongruity.*   When asked what, in all his travels, was the strangest thing he had ever seen, Thales replied, "An aged tyrant."

31° *The Thales puzzle.*   It has been reported that Thales resided for a time in Egypt, and there evoked admiration by calculating the height of a pyramid by means of shadows. Two versions of the story have been given. The earlier account, furnished by Hieronymus, a pupil of Aristotle, says that Thales noted the length of the shadow of the pyramid at the moment when his own shadow was the same length as himself. The later version, given by Plutarch, says that Thales set up a stick and employed the fact that "the [height of the] pyramid was to the [length of the] stick as the shadow of the pyramid to the shadow of the stick." Both of these versions fail to mention the very real difficulty, in either case, of obtaining the length of the shadow of the pyramid (that is, the distance from the apex of the shadow to the center of the base of the pyramid).

The above unaccounted-for difficulty has given rise to what might be called the *Thales puzzle*: Devise a method, based on shadow observations and similar triangles and independent of latitude and specific time of day or year, for determining the height of a pyramid. Figure 4

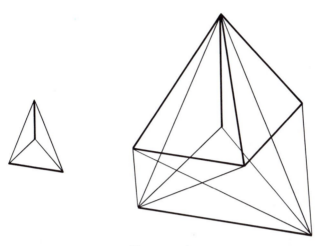

FIGURE 4

suggests a solution—that the reader may care to complete—requiring *two* shadow observations spaced a few hours apart.

32° *Thales, the engineer.* Thales was a resourceful man and has been credited with a now doubted spectacular feat in engineering. King Croesus, who was an admirer and probably one-time patron of Thales, had urgent need to transport his army across the river Halys. Boats were out of the question, pontoon bridges were still a thing of the future, and there was no time to build a permanent bridge. To resolve his predicament, Croesus consulted Thales. The wise man quickly saw a solution. He directed that a canal be excavated to divert the river into a temporary channel. After the army had passed dryshod over the former river bed, Thales had everything restored to its former order, so as not to offend the river gods that Croesus respected.

33° *Thales, the astronomer.* There is another story, also doubted today, that Thales predicted the total, or nearly total, eclipse of the sun that took place on May 28, 585 B.C. According to one version of the story, on that day the Medes and Lydians were locked in a battle during the sixth year of a stubborn war. The soldiers suddenly found themselves fighting in growing darkness. Terrified by the vanishing of the sun, the warring hosts quit the fight. A peace was concluded and later sealed by a double marriage between their respective reigning families.

A precise prediction of the time and locality of an eclipse of the sun requires a mathematical finesse far beyond what was possible in Thales' time. To have predicted just the year and the rough quarter of the globe of such an occurrence would have been a praiseworthy accomplishment in those days.

Also in connection with astronomy, it is said that Thales advised mariners to steer by the Little Bear rather than the Great Bear.

34° *Thales, the statesman.*   Among the many stories subsequently told of Thales by his admirers is that of his simple solution of the incessant brawling among the five Ionian city-states and of their openness as prey to outside invaders. The internecine warfare stopped and joint safety was secured when Thales suggested a federation of the city-states.

PYTHAGORAS

THE next man mentioned in the history of mathematics, after the illustrious Thales, is Pythagoras, who became enveloped by his followers in such a mythical haze that very little is known about him with any degree of certainty. After extensive traveling, he settled in the Greek seaport of Crotona, located in southern Italy. There he founded the famous Pythagorean School, which, in addition to being an academy for the study of philosophy, mathematics, and natural science, developed into a closely knit brotherhood with secret rites and observances. Many stories have come down to us about Pythagoras and his Brotherhood.

35° *The lure of geometry.*   It has been told that Pythagoras, in an effort to secure some students, found an impecunious young artisan and offered to teach him geometry. To make the lessons worth the poor pupil's time, the master agreed to pay him a penny for each geometrical theorem he mastered. This suited the young man well, for he found that by listening attentively he could earn better wages in an hour than in a full day at his usual labor. As his pile of pennies grew, the pupil discovered that in spite of himself he was becoming more interested in his studies than in his increasing wealth. Indeed, so avid

did his interest in geometry become and so thoroughly did he fall under
the subject's spell, that he begged his teacher to proceed faster, and as
an inducement insisted that Pythagoras now accept from him a penny
for each new theorem. By the time the pupil absorbed all the geometry
he could hold, the gratified teacher had gained back all of his pennies.

36° *The first recorded facts in mathematical physics.* A remarkable discovery about numbers, claimed to have been made by Pythagoras, is the dependence of musical intervals upon simple numerical
ratios. Thus Pythagoras is said to have found that for like strings under
the same tension, the lengths should be 2 to 1 for the octave, 3 to 2 for
the fifth, and 4 to 3 for the fourth. These results, apparently the first
recorded facts in mathematical physics, are said to have led the
Pythagoreans to initiate the scientific study of musical scales.

The manner of the above discovery by Pythagoras has been told
repeatedly over the ages and runs as follows. While passing a blacksmith's shop one day, Pythagoras was arrested by the clang of four
hammers swung in succession by four slaves pounding a piece of red-hot
iron. All but one of the hammers clanged in harmony. Upon investigation, Pythagoras found that the differences in pitch of the four sounds
were due to the different weights of the hammers. He persuaded the
blacksmith to let him borrow the hammers for a short time. Taking
the hammers home, he weighed them and then hung them individually
to four strings of the same length and thickness. On plucking the
strings, he noted that the emitted sounds corresponded to those made
by the hammers striking the red-hot iron. By sticking a small lump of
clay on the hammer responsible for the dissonance, he brought the
emitted note for this hammer into harmony with the other three.

Now, curiously enough, there is a grave flaw in this often-repeated
story, namely: *the tone of a hammer striking a given anvil is independent of the
weight of the hammer.* A couple of minutes in a blacksmith's shop would
have convinced any one of the erudite scholars who passed this legend
on to their successors that the whole story is physically absurd.

37° *A hecatomb of oxen.* Legend has it that Pythagoras celebrated his discovery of the famous right-triangle theorem by sacrificing
to the gods a hecatomb (that is, 100 head) of oxen. Concerning this

sacrifice, C. L. Dodgson (Lewis Carroll), in his *A New Theory of Parallels* published in 1895, says:

> But neither thirty years, nor thirty centuries, affect the clearness, or the charm, of Geometrical truths. Such a theorem as "the square of the hypotenuse of a right-angled triangle is equal to the sum of the squares of the sides" is as dazzlingly beautiful now as it was in the day when Pythagoras first discovered it, and celebrated its advent, it is said, by sacrificing a hecatomb of oxen—a method of doing honor to Science that has always seemed to me *slightly* exaggerated and uncalled-for. One can imagine oneself, even in these degenerate days, marking the epoch of some brilliant scientific discovery by inviting a convivial friend or two, to join one in a beefsteak and a bottle of wine. But a *hecatomb* of oxen! It would produce a quite inconvenient supply of beef.

38° *A play on words.* In the German vernacular a dunce or blockhead is called an ox, or *Ochs*. Now after Pythagoras discovered his famous theorem he sacrificed a hecatomb of oxen. Since that time all *Ochsen* tremble whenever a new truth is discovered.

39° *A philosopher.* If the following sentiments, attributed to Pythagoras, are authentic, they furnish a fine summary of his life, character, and ideals.

"I have no trade," he once declared; "I am a philosopher."

"And what may that be?" he was asked.

"This life," he explained, "may be compared to the Olympic games. For in that concourse some seek glory or strive for wreaths; others, peddling goods, pursue profit; others again, less base than either, go to the games neither for applause nor for gain, but merely to enjoy the sport and keep abreast of the times.

"In the same way we men quitted our celestial home and came into this world, where many toil for honor and the majority for gain, and where but a few, despising greed and vanity, study nature for its own sake. These last I call philosophers."

40° *Friendship.* Asked what a friend is, Pythagoras replied, "Another I." From this developed the amicability of numbers. What could depict a closer friendship than the *amicable number pair* 220 and

284? The proper divisors* of 220 are 1, 2, 4, 5, 10, 11, 20, 22, 44, 55, 110, and the sum of these is 284; the proper divisors of 284 are 1, 2, 4, 71, 142, and the sum of these is 220. Each of the two numbers 220 and 284 generates the other; surely nothing can be more intimate than this.

41° *The marriage of Pythagoras.* More than one teacher has married a young and admiring pupil. Pythagoras may have been the first to do this. According to the story, among Pythagoras's favorite pupils was Theano, the beautiful young daughter of his host Milo. Theano developed a hopeless infatuation for her teacher, who was so deeply submerged in his studies that he noticed nothing. So he was greatly surprised one day when Theano informed him that she was about to expire of an unreciprocated passion for someone. On persistent questioning by the master, Theano finally admitted that Pythagoras himself was the man she loved. To save her sanity, if not her life, Pythagoras sacrificed his asceticism and married her.

42° *Pythagorean teaching.* Pythagoras taught his Brothers to refrain from wearing wool clothing (because of the transmigration of souls of men into animals), never to take a higher road if a lower one should be present (because of humility), and never to poke a fire with iron (because flame was the symbol of truth). More obscurely, he instructed the Brothers not to sit on a quart measure, not to touch a white rooster, and not to eat beans. He also taught the Brothers celebacy—in spite of his marriage to Theano.

43° *Pythagoras's golden thigh.* The misanthropic philosopher Heraclitus has said that Pythagoras was the son of Mnesarchus, a stone cutter of Samos. And there is a vague rumor that Pythagoras's mother was of Phoenician extraction, and that she accompanied her son on one of his journeys. But to Pythagoras's disciples, the master was of divine origin, having for heavenly father none other than Apollo. In proof of this celestial descent, Pythagoras is said to have possessed a golden thigh. This curious legend is so persistent that it has been

---

* The *proper divisors* of a positive integer N are all the positive integral divisors of N except N itself. Note that 1 is a proper divisor of N. A somewhat antiquated synonym for proper divisor is *aliquot part.*

wondered if it might be a miracle-monger's distortion of some real physical infirmity, and, if so, what the infirmity might have been.

44° *The end of Pythagoras.* In time the influence and aristocratic tendencies of the Pythagorean Brotherhood in Crotona became so great that the democratic forces of southern Italy destroyed the buildings of the school and caused the society to disperse. Pythagoras fled to Metapontum where he died, or was murdered, at an advanced age of 75 to 80. According to one legend, the master, fleeing from his pursuers, was forced into a bean field. Rather than chance treading on one of the sacred bean plants, Pythagoras chose death.

45° *Pythagoras's proof of his theorem.* Tradition is unanimous in ascribing to Pythagoras the independent discovery of the theorem on the right triangle that now universally bears his name, that the square on the hypotenuse of a right triangle is equal to the sum of the squares on the two legs. We have seen, in Item 17°, that this theorem was known to the Babylonians of Hammurabi's time, more than a thousand years earlier, but the first general proof of the theorem may well have been given by Pythagoras. There has been much conjecture as to the proof Pythagoras might have offered, and it is generally felt that it probably was a dissection type of proof* like the following, illustrated in Figure 5. Let a, b, c denote the legs and hypotenuse of the

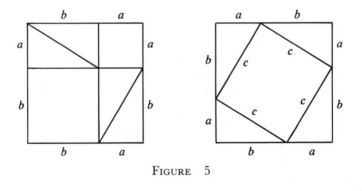

FIGURE 5

---

* See, however, Daniel Shanks, *Solved and Unsolved Problems in Number Theory*, Vol. 1 (Washington, D.C.: Spartan Books, 1962), pp. 124, 125.

given right triangle, and consider the two squares in the figure, each having $a + b$ as side. The first square is dissected into six pieces, namely the two squares on the legs and four right triangles congruent to the given triangle. The second square is dissected into five pieces, namely the square on the hypotenuse and again four right triangles congruent to the given triangle. By subtracting equals from equals, it now follows that the square on the hypotenuse is equal to the sum of the squares on the legs.

To prove that the central piece of the second dissection is actually a square of side c we need to employ the fact that the sum of the angles of a right triangle is equal to two right angles. But the Greek commentator Proclus attributes this theorem for the general triangle to the Pythagoreans. Since a proof of this theorem requires, in turn, a knowledge of some properties of parallels, the early Pythagoreans are also credited with the development of that theory.

THE PYTHAGOREAN BROTHERHOOD

46° *Motto of the Pythagorean Brotherhood.* The Pythagorean attitude of knowledge for its own sake, rather than for profit, is illustrated by the motto of the Brotherhood:

> A figure and a step onward;
> Not a figure and a florin.

47° *Himself said it.* The early Pythagoreans so revered the founder of their Brotherhood that it was considered impious for a member to claim any discovery for his own glory, but all must be referred back to the master himself. It is told that the Pythagorean Hippasus was justly drowned at sea in a shipwreck as punishment for irreverently claiming the discovery of the regular dodecahedron.

48° *Brotherhood loyalty.* A young and penniless member of the Pythagorean Brotherhood fell desperately ill while traveling in foreign lands and was taken to the nearest inn. There the charitable innkeeper nursed him, even though the young man made it clear that he had neither money nor goods with which to repay his host. When it became

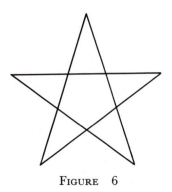

FIGURE 6

certain that he was dying, the young man asked the innkeeper for a board on which to draw. With effort he managed to scrawl on the board the mystic pentagram symbol (see Figure 6) of his Brotherhood. Turning to the innkeeper, he said, "My good friend, hang this board outside your inn door. Some day a traveler who understands what I have drawn will pass this way and will stop and ask you about the sign. Tell him everything, and you will be rewarded." And just so did it happen.

49° *Damon and Phintias.* There were two Syracusan youths of the fourth century B.C. who followed the Pythagorean mandate concerning friendship to such a degree that their names have become synonymous with the fine loyalty of that relationship. Phintias (sometimes incorrectly called Pythias) was sentenced by the tyrant Dionysius of Syracuse to die for treason. The condemned man begged a postponement of his sentence so that he might first return to his home in a neighboring village to arrange his affairs. His petition was granted provided he could obtain someone to remain in his cell as a pledge of his return. His friend Damon offered to do this, and expressed a willingness to die in Phintias's stead should Phintias not appear by the fixed time of the postponed execution. Unexpected predicaments delayed Phintias, and the scheduled day arrived. Damon, still fully believing in his friend's faithfulness, was led out for the execution. People gathered and began to express pity at Damon's credulity, when suddenly the delayed Phintias rushed breathlessly through the crowd into the arms of his friend. Each youth demanded to die for the other. Struck by this loyalty of friendship, Dionysius released both youths and expressed a desire to be admitted to their friendship.

50° *The three questions.* Every evening each member of the Pythagorean Brotherhood had to put three questions to himself: In what have I failed? What good have I done? What have I not done that I ought to have done?

PYTHAGOREANISM

51° *The transmigration of souls.* Pythagoras taught the transmigration of the individual soul from one body to another, even to that of a different species. Although bodies served as tombs or prisons of the soul, it was admitted that if a man led a sufficiently pure life his soul might be released from all flesh.

There is a Pythagorean legend that tells of a Brother (or perhaps Pythagoras himself) coming upon a citizen of Crotona beating his dog. The Brother charged the man to stop, declaring that in the dog's yowls of pain he recognized the cries of a departed friend pleading for mercy. "For the very sin you are committing," the Brother informed the citizen, "my friend is now this dog with you as his harsh master. At the next turn of the Wheel of Birth, he may be the master and you the dog. Pray that he will be more merciful to you than you are to him. It is only thus that he can hope to escape the Wheel." That the Pythagorean School was respected is evidenced by the fact that the legend claims the citizen immediately stopped beating the dog and begged the animal's forgiveness.

52° *Number rules the universe.* The Pythagoreans believed that the whole numbers were the elements of everything. Even such concepts as reason, justice, man, health, and marriage were identified with distinct whole numbers. Thus the whole numbers ruled the universe, and therefore one who should master the relationships of the whole numbers might be able to understand and perhaps even guide the affairs of the universe.

The lower whole numbers had the following special meanings.

One, as the originator of all numbers, was the most revered digit and was the number of reason. Two, the first even or female number, represented diversity of opinion. Three, the first true male number, represented harmony, being composed of unity and diversity. Four stood for justice or retribution, suggesting the squaring of accounts. Five, being the union of the first true male and female numbers, represented marriage. Six was the number of creation.

In their diligent effort to search out the relationships of the whole numbers, the Pythagoreans took the first steps in the development of number theory, and at the same time laid much of the basis of future number mysticism.

53° *Amicable numbers.* Two whole numbers, it will be recalled from Item 40°, are *amicable* if each is the sum of the proper divisors of the other. For example, 284 and 220, which constitute a pair that has been ascribed to Pythagoras, are amicable. This particular pair of numbers attained a mystical aura, and superstition maintained that two talismans bearing these numbers would seal perfect friendship between the wearers; should one of the wearers, though separated from the other by half the circumference of the earth, receive even so small an injury as a pin prick, the other would be aware of it. The numbers came to play an important role in magic, sorcery, astrology, and the casting of horoscopes.

Curiously enough, it seems that no new pair of amicable numbers was discovered until the great French number theorist Pierre de Fermat in 1636 announced 17,296 and 18,416 as another pair. Two years later the French mathematician and philosopher René Descartes gave a third pair. The Swiss mathematician Leonard Euler undertook a systematic search for amicable numbers and, in 1747, gave a list of thirty pairs, which he later extended to more than sixty. A second curiosity in the history of these numbers was the late discovery, by the sixteen-year-old Italian boy Nicolo Paganini in 1866, of the overlooked and relatively small pair of amicable numbers, 1184 and 1210. Today more than nine hundred pairs of amicable numbers are known. These pairs are all of the same parity; that is, the two numbers of a pair are either both odd or both even. All the odd ones are multiples of 3; all the even pairs have the sum of their digits a multiple of 9.

A sequence of three or more numbers such that adjacent pairs of the numbers are amicable is known as a *sociable chain* of numbers. Only two sociable chains are known. In 1918 the Frenchman P. Poulet found one of five "links":

12,496,   14,288,   15,472,   14,536,   14,264.

There is a twenty-eight-link chain starting with 14,316. A sociable chain of exactly three links has been called a *crowd*; no crowds have yet been found.

54° *Deficient, perfect, and abundant numbers.* Other numbers having mystical connections essential to numerological speculations, and sometimes ascribed to the Pythagoreans, are the *perfect, deficient,* and *abundant numbers.* A number is *perfect* if it is the sum of its proper divisors, *deficient* if it exceeds the sum of its proper divisors, and *abundant* if it is less than the sum of its proper divisors. So God created the world in six days, a perfect number, since $6 = 1 + 2 + 3$. On the other hand, as Alcuin (735–804) observed, the whole human race descended from the eight souls of Noah's ark, and this second creation was imperfect, for 8, being greater than $1 + 2 + 4$, is deficient. And thus we account for the many ills of our present world.

Until 1952 there were only twelve known perfect numbers, all of them even numbers, of which the first three are 6, 28, and 496. The last proposition of the ninth book of Euclid's *Elements* (ca. 300 B.C.) proves that *if $2^n - 1$ is a prime number, then $2^{n-1}(2^n - 1)$ is a perfect number.* The perfect numbers given by Euclid's formula are even numbers, and Euler has shown that every even perfect number must be of this form. The existence or nonexistence of odd perfect numbers is one of the celebrated unsolved problems in number theory. There certainly is no number of this type having less than twenty digits.

In 1952, with the aid of the SWAC digital computer, five more perfect numbers were discovered, corresponding to $n = 521, 607, 1279$, 2203, and 2281 in Euclid's formula. In 1957 the Swedish machine BESK found another, corresponding to $n = 3217$, and in 1961 an IBM 7090 found two more, for $n = 4253$ and 4423. There are no other perfect numbers for $n < 5000$.

The values $n = 9689, 9941$, and 11,213 also yield perfect numbers, bringing the list of known perfect numbers to 23. The perfect number corresponding to $n = 11,213$ was found in 1963 at the University of Illinois. This very large number consists of 6751 digits and has 22,425 divisors. The University of Illinois mathematics department is so proud of the discovery of this largest known perfect number that its postage meter has been stamping on envelopes a rectangle bearing the statement, "$2^{11213} - 1$ is prime."

The pursuit of larger and larger perfect numbers shows how wrong was Peter Barlow who, in his *Theory of Numbers* of 1811, wrote about the ninth perfect number (corresponding to $n = 61$): "It is the greatest that will be discovered, for, as they are merely curious without being useful, it is not likely that any person will attempt to find one beyond it."

There are only twenty-one abundant numbers between 10 and 100, and these are all even. That all abundant numbers are not even follows from the easily established fact that $945 = 3^3 \cdot 5 \cdot 7$ is abundant. This is the first odd abundant number, and the only odd abundant number not exceeding 1000.

55° *Pythagorean philosophy and geometry at stake.* Pythagoras preached, with all the fervor of a Savonarola, that everything—simply everything—depends upon the whole numbers. Imagine the consternation, then, within the Pythagorean ranks when some Brother discovered the devastating fact that $\sqrt{2}$ is not a rational number, that is, cannot be expressed as the ratio of two whole numbers, and therefore appears to be independent of the whole numbers. The geometrical counterpart of this discovery was equally startling, for if $\sqrt{2}$ is not a rational number, it follows that a side and diagonal of a square possess no common unit of measure that can be stepped off exactly a whole number of times into each. This contradicted the firm intuitive belief that any two line segments must possess some common unit of measure, though perhaps very, very small. Now the entire Pythagorean theory of proportion and similar figures was built upon this seemingly obvious assumption. In one fell stroke, both the basic Pythagorean philosophy and much of Pythagorean accomplishment in geometry were threatened. It is reported that so great was the "logical scandal" that efforts were made for a while to keep the matter secret, and one legend has it that the first Pythagorean to divulge the secret to outsiders was banished from the Brotherhood and a tomb was erected for him as though he were dead.

56° *Pythagoras justified.* Every student of college mathematics learns of the remarkable mathematical accomplishment made toward the end of the nineteenth century wherein, by starting from a postulational development of the whole numbers (that is, positive integers) and making no further assumptions, one first obtains the set of all integers, then the set of all rational numbers, and then the set of all real numbers. Since the irrational numbers, like $\sqrt{2}$, are among the real numbers, we see that, at least so far as the real number system is concerned, the ancient Pythagorean belief that everything depends upon the whole numbers is today justified.

57° *The case for Pythagoreanism.* Pythagoras said that number is everything, but, aside from his analysis of musical tones (see Item 36°), he did not make a strong case for his assertion. He could not be expected to do so, for science in his time was at a very primitive level. However, if we should ask whether modern physical scientists believe that the world can best be understood numerically, the answer is an overwhelming yes.

An examination of a list of the key discoveries in physical science shows that a very strong case can be built for the Pythagorean concept that number rules the universe. Just consider Galileo's law of falling bodies, Kepler's laws of planetary motion, Newton's law of universal gravitation, Coulomb's law, Maxwell's electromagnetic wave equations, Prout's law of definite proportion and Dalton's law of multiple proportion in chemistry, Lisle's law of constant angles and Haüy's law of rational indices in crystallography, the Dulong–Petit law for the specific heats of solids, Faraday's law of electrolysis, the optical spectrum resulting from Fraunhofer's diffraction grating, Stefan's law of radiation, Planck's quanta, Mendeléeff's periodic chart of the chemical elements, Schroedinger's wave equation, and so on and on and on.

PLATO

Plato, "the most winged, most charming, and best loved of all the philosophers of the Western World," was born in or near Athens in 427 B.C. He studied philosophy under Socrates there, and then set out upon his extensive wanderings for wisdom, studying mathematics under Theodorus of Cyrene on the African coast and becoming an intimate friend of the eminent Pythagorean Archytas. Upon his return to Athens around 387 B.C., he founded his famous Academy there, an institution for the systematic pursuit of philosophical and scientific inquiry. He presided over his Academy for the rest of his life, and died in Athens in 347 B.C. at the venerable age of eighty. Almost all the important mathematical work of the fourth century B.C. was done by friends or pupils of Plato, making his Academy the link between the mathematics of the earlier Pythagoreans and that of the long-lived school of mathematics at Alexandria. The Athenian school lasted some nine hundred years, until the Christians obtained from Emperor Justinian a decree that in 529 A.D. closed its doors forever.

58° *Plato's motto and the transfer of training.* Plato's influence on mathematics was not due to any mathematical discoveries he made, but rather to his enthusiastic conviction that the study of mathematics furnished the finest training field for the mind, and hence was essential in the cultivation of philosophers and those who should govern his ideal state. This belief explains the renowned motto over the door of his Academy:

> Let no one unversed in geometry enter here.

Thus, because of its logical element and the pure attitude of mind that he felt its study creates, mathematics seemed of utmost importance to Plato, and for this reason it occupied a valued place in the curriculum of the Academy.

In recent decades there has been vigorous controversy over the question of mental discipline and transfer of training, with contentions ranging from one extreme to the other. Thus there are those who, like Plato, claim that the study of mathematics develops the pupil's respect for truth and therefore results in honesty, that it develops neatness, power of concentration, and especially the power to think clearly. But the advent of a mechanistic psychology cast grave doubt upon the validity of such claims, and it was quite conclusively shown that transfer of training is not complete, automatic, and inevitable. There are many examples of dishonesty, lack of neatness, and failure to think clearly in nonmathematical situations on the part of students of mathematics, and many uncritical people have interpreted this to mean that transfer of training is nonexistent. Competent psychologists today seem agreed that the truth lies between the two extreme positions; transfer of training *can* take place *if* the subject is taught with this specific purpose in view.

59° *Michel Chasles and the forged autograph letters.* [The following true story, which is related to Item 58° above, is a mixture of the amusing, the pathetic, and the incredible. It is here told, with permission, in the words of Dean R. A. Rosenbaum, from his elegant narration in the Historically Speaking section of *The Mathematics Teacher*, May, 1959, pp. 365–366.]

Educators often consider the question of "transfer." For example, does the study of mathematics instill habits of logical thought which carry over to the analysis of problems in other rational disciplines? It is of interest in this connection to examine the behavior of professional mathematicians, to see whether they exhibit notably logical qualities in the nonmathematical aspects of their lives. The spectacle of a mathematician acting in a markedly irrational manner fills most observers with unholy glee. No such spectacle can surpass that of Michel Chasles and the forged autograph letters.

Chasles was one of the foremost geometers of the nineteenth century. His *Aperçu historique sur l'origine et le développement des méthodes en Géométrie . . .*, published as a memoir of the Academy of Brussels in 1837, is an extraordinary achievement of synthesis and generalization which won him immediate recognition. He contributed many theorems to geometry, and the "principle of algebraic correspondence" is known by his name. Joseph Bertrand quotes what he refers to as an oft-repeated sentence, "All the geometers of Europe are disciples of M. Chasles."

But Chasles was an especially ardent French patriot, and his

nationalistic pride led to a debacle. When shown some letters, pur-
portedly written by Pascal, in which the laws of gravitational attraction
were set out, he eagerly bought them; here was proof of France's
priority to Newton's England! A scholar and bibliophile with a
comfortable income, Chasles continued to buy documents from one
Vrain-Denis Lucas during the period 1861–69.

The details of his purchases seem incredible. He bought over
27,000 letters, for about 140,000 francs. There were 175 letters from
Pascal to Newton, 139 from Pascal to Galileo, and large numbers
written by Galileo. But Lucas provided ancient, nonmathematical
letters as well. Included in Chasles' purchases were six from Alexander
the Great to Aristotle, one from Cleopatra to Caesar, one from Mary
Magdalene to Lazarus, and one from Lazarus to St. Peter. Every
letter was written on paper, and in French! It is probably true that
Chasles, in his ardor and enthusiasm, did not look at many of his
27,000 purchases.

When Chasles disclosed to the French Academy of Sciences his
theory of Pascal's priority to Newton, there was considerable scepticism.
Chasles displayed some of his letters, and it was pointed out that the
handwriting was not the same as that of letters which were indubitably
Pascal's. Various anachronisms appeared. Each was met by a new
letter furnished by Lucas, in which the difficulties were explained away.
But after several years of controversy, Chasles had to acknowledge
defeat. He exhibited his entire stock of 27,000 forged letters, and Lucas
was sent to prison for two years.

Lucas's defense at his trial was interesting. He maintained that he
had done nothing wrong—that Chasles had really received his money's
worth, that the controversy and trial, which were widely reported, had
stimulated in the public a healthy interest in history, that the debates
in the Academy had been much more exciting than usual, and that he,
himself, had acted through patriotic motives.

At the same time that one marvels at Chasles' gullibility, one
must be amazed by Lucas's industry: to "antique" paper for 27,000
letters is itself quite a task! Lucas apparently had spent many hours
each day in libraries, acquiring historical knowledge for his writings.
Since he knew neither Greek nor Latin, he was severely handicapped
in his work. According to J. A. Farrer (in his *Literary Forgeries*, Long-

mans, Green, and Co., 1907, Chap. xii), nothing is known of Lucas after his prison sentence; but Bertrand reports that Lucas served his time, returned to his "profession" after release, and was resentenced to three years as a recidivist, Chasles asking wryly, "Wouldn't it have been better to sentence him to five years from the start?"

Apparently Chasles was greatly embarrassed by the affair, and Bertrand remarks that the man had suffered enough—the matter should be forgotten. But Farrer can't resist giving a twist of the knife, which may well serve as a warning to all of us:

> The logical incapacity that M. Chasles displayed throughout the contest subsequently waged over his supposed treasures shows conclusively how insignificant is the benefit conferred on the reasoning faculties by mathematical studies. The leading mathematician of his country showed himself incapable of reasoning better than a child.

60° *Some particularly elusive Platonic numerology.* Plato was much influenced by the early Pythagoreans and it would seem that, at heart, he was a numerologist so far as his own mathematical beliefs went. There is no profundity in all of his works that has given his commentators more trouble than the following passage from Book VIII of his *Republic*:

> For that which, though created, is divine, a recurring period exists, which is embraced by a perfect number. For that which is human, however, by that one for which it first occurs that the increasings of the dominant and the dominated, when they take three spaces and four boundaries making similar and dissimilar and increasing and decreasing, cause all to appear familiar and expressible.
>
> Whose base, modified, as four to three, then married to five, three times increased, yields two harmonies: one equal multiplied by equal, a hundred times the same: the other equal in length to the former, but oblong, a hundred of the numbers upon expressible diameters of five, each diminished by one, or by two if inexpressible, and a hundred cubes of three. This sum now, a geometrical number, is lord over all these affairs, over better and worse births; and when in ignorance of them, the guardians unite the brides and bridegrooms wrongly, the children will not be well-endowed, either in their constitutions or in their fates.

The reader may care to try to unravel the mathematical basis of this strange and oracular passage. The passage defied analysis even by

Plato's immediate successors, and it was not until some 2300 years later that a satisfactory explanation was given. The elucidation appeared in a fascinating article in the *Proceedings of the London Mathematical Society* of 1923, and was given by Grace Chisholm Young, an eminent British mathematician of the early twentieth century. One must read Mrs. Young's article to appreciate her penetrating analysis, but one remarkable conclusion reached by Mrs. Young is that Plato guessed, and possibly proved, that *the only relatively prime whole numbers x, y, z, w simultaneously satisfying*

$$x^2 + y^2 = z^2 \quad and \quad x^3 + y^3 + z^3 = w^3,$$

are x = 3, y = 4, z = 5, w = 6.

61° *The five Platonic solids.*   A polyhedron is said to be *regular* if its faces are congruent regular polygons and its polyhedral angles are all congruent. While there are regular polygons of all orders, it is surprising that there are only five different regular polyhedra. The regular polyhedra are named according to the number of faces each possesses. Thus there is the tetrahedron with 4 triangular faces, the hexahedron (or cube) with 6 square faces, the octahedron with 8 triangular faces, the dodecahedron with 12 pentagonal faces, and the icosahedron with 20 triangular faces. See Figure 7.

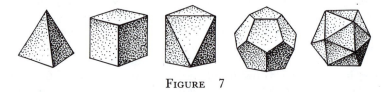

FIGURE   7

The early history of these regular polyhedra is lost in the dimness of the past. A mathematical treatment of them is initiated in Book XIII of Euclid's *Elements*. The first scholium (in all likelihood added later and probably taken from Geminus) of this book remarks that the book "will treat of the so-called Platonic solids, incorrectly named, because three of them, the tetrahedron, cube, and dodecahedron are due to the Pythagoreans, while the octahedron and icosahedron are due to Theaetetus." This could well be the case.

In any event, a description of all five regular polyhedra was given by Plato, who, in his *Timaeus*, shows how to construct models of the solids by putting triangles, squares, and pentagons together to form their faces. Plato's Timaeus is the Pythagorean Timaeus of Locri, whom Plato presumably met when he visited Italy. In Plato's work, Timaeus mystically associates the four easily constructed solids—the tetrahedron, octahedron, icosahedron, and cube—with the four Empedoclean primal "elements" of all material bodies—fire, air, water, and earth. The disturbing difficulty of accounting for the fifth solid, the dodecahedron, is taken care of by associating it with the enveloping universe.

62° *Kepler's explanation of the Timaeus associations.* Johannes Kepler, master astronomer, mathematician, and numerologist of later times (1571–1630), gave an ingenious explanation of the Timaeus associations of the five Platonic solids. Of the regular solids, he intuitively assumed that the tetrahedron encloses the smallest volume for its surface, while the icosahedron encloses the largest. Now these volume–surface relations are qualities of dryness and wetness, respectively, and since fire is the driest of the four "elements" and water the wettest, the tetrahedron must represent fire and the icosahedron water. The cube is associated with earth since the cube, resting foursquarely on one of its square faces, has the greatest stability. On the other hand, the octahedron, held lightly by two of its opposite vertices between a forefinger and thumb, easily spins and has the instability of air. Finally, the dodecahedron is associated with the universe because the dodecahedron has twelve faces and the zodiac has twelve signs.

63° *The Platonic solids in nature.* The tetrahedron, cube, and octahedron can be found in nature as crystals, for example as crystals of sodium sulphantimoniate, common salt, and chrome alum, respectively. The other two cannot occur in crystal form, but have been observed as skeletons of microscopic sea animals called *radiolaria*.

64° *The most extraordinary application of the Platonic solids to a scientific problem.* Kepler's mystical awe of the five Platonic solids extended beyond the association of these solids with the four primal "elements" and the enveloping universe. According to Kepler, these five solids accounted for both the number of planets (only five were known in his day) and the way the planets are spaced about the sun. In his *Mysterium cosmographicum* of 1596, Kepler says:

> The orbit of the Earth is a circle: round the sphere to which this circle belongs, describe a dodecahedron; the sphere including this will give the orbit of Mars. Round Mars describe a tetrahedron; the circle including this will be the orbit of Jupiter. Describe a cube round Jupiter's orbit; the circle including this will be the orbit of Saturn. Now inscribe in the Earth's orbit an icosahedron; the circle inscribed in it will be the orbit of Venus. Inscribe an octahedron in the orbit of Venus; the circle inscribed in it will be Mercury's orbit. This is the reason of the number of the planets.

We see that Johannes Kepler was a confirmed Pythagorean. He even once suggested the possibility that the soul of Pythagoras may have taken up residence in his body.

65° *Some problems concerning the Platonic solids.*

(1) In Item 61°, the definition of regularity of a polyhedron involves three properties: regular faces, congruent faces, congruent polyhedral angles. Many textbooks on solid geometry do not give all three of the defining properties. The interested reader may care to show, by counterexamples, that all three properties are necessary.

(2) From the three defining properties listed in (1), one can deduce the regularity of the polyhedral angles. The reader is invited to do this, and then to show that the three defining properties can be replaced by only two: regular faces and regular polyhedral angles.

(3) The uninitiated will almost always intuitively believe that of the regular dodecahedron (a solid having 12 faces) and a regular icosahedron (a solid having 20 faces) inscribed in the same sphere, the icosahedron has the greater volume. The reader is invited to show that the reverse is actually the case, and also to show that of a cube (a solid having 6 faces) and a regular octahedron (a solid having 8 faces) inscribed in the same sphere, the cube has the larger volume.

(4) It is interesting that a regular dodecahedron and a regular icosahedron inscribed in the same sphere have a common inscribed

sphere. Prove this, and show that the same is true of a cube and a regular octahedron inscribed in the same sphere.

(5) In Item 62° we noted that Kepler intuitively assumed that, of the five regular solids, for a given surface area the tetrahedron encloses the smallest volume and the icosahedron encloses the largest volume. Is this so?

(6) A regular dodecahedron, a regular icosahedron, and a cube are inscribed in the same sphere. Prove that the volume of the dodecahedron is to the volume of the icosahedron as the length of an edge of the cube is to the length of an edge of the icosahedron.

(7) In the so-called Book XIV of Euclid's *Elements*, it is pointed out that Hypsicles observed that if a regular dodecahedron and a regular icosahedron are inscribed in the same sphere, then their volumes are in the same ratio as their surface areas. Prove this.

(8) Show that the circumcircles of the faces of the regular dodecahedron and the regular icosahedron inscribed in the same sphere are equal.

66° *The Delian problem.* Plato was born in, or close to, the year of the plague that killed a large portion of the population of Athens. The great impression that this catastrophe made may be the origin of a famous mathematical problem of Greek antiquity.* There is a report that a delegation was sent to the oracle of Apollo on the island of Delos to inquire how the plague might be averted. The oracle replied that the cubical altar to Apollo must be doubled in size. The Athenians accordingly doubled the dimensions of the altar, but the plague did not subside. Returning to the oracle for an explanation of the failure, the Athenians were informed that they had not carried out the orders; they had not *doubled* the size of the altar, but had increased it to *eight times* its original size. Thus arose the geometry problem: to find the edge of a cube whose volume shall be exactly twice the volume of a given cube. The problem remained refractory and in time was reputedly taken to Plato, who submitted it to the geometers. It is this story that has led the problem of duplicating a cube frequently to be referred to as the *Delian problem.* Whether the story is true or not, the problem was

---

* For a more likely origin of the problem, see Item 246°.

studied in Plato's Academy, and there are higher geometry solutions attributed to Eudoxus, to Menaechmus, and even (though probably erroneously) to Plato himself.

EUCLID

DISAPPOINTINGLY little is known about the life and personality of Euclid except that he was the first professor of mathematics at the famed Museum of Alexandria (which opened its doors about 300 B.C.), and apparently the founder of the illustrious and long-lived Alexandrian School of Mathematics. Even his dates and his birthplace are not known, but it seems probable that he received his mathematical training in the Platonic school at Athens. Although he was the author of at least ten works, and fairly complete texts of five of these have come down to us, his reputation rests mainly on his *Elements*. As soon as this work appeared it was accorded the highest respect. No work, except the Bible, has been more widely used, edited, or studied, and probably no work has exercised greater influence on scientific thinking.

67° *The royal road in geometry.* Only two anecdotes about Euclid have come down to us, and both are doubtful. In his *Eudemian Summary*, Proclus (410–485) tells us that Ptolemy Soter, the first King of Egypt and the founder of the Alexandrian Museum, patronized the Museum by studying geometry there under Euclid. He found the subject difficult and one day asked his teacher if there weren't some easier way to learn the material. To this Euclid replied, "Oh King, in the real world there are two kinds of roads, roads for the common people to travel upon and roads reserved for the King to travel upon. In geometry there is no royal road."

This is an example of an anecdote told also in relation to other people, for Stobaeus has narrated it in connection with Menaechmus when serving as instructor to Alexander the Great.

Since so many students are considerably more able as algebraists than as geometers, analytic geometry, which studies geometry with the aid of algebra, has been described as the "royal road in geometry" that Euclid thought did not exist.

68° *Euclid and the student.* The second anecdote about Euclid that has come down to us is an unreliable but pretty story told by Stobaeus in his collection of extracts, sayings, and precepts for his son. One of Euclid's students, when he had learned the first proposition, asked his teacher, "But what is the good of this and what shall I get by learning these things?" Thereupon Euclid called a slave and said, "Give this fellow a penny, since he must make gain from what he learns."

69° *Euclid's* Elements *compared with* Newton's Principia. Augustus De Morgan (1806–1871) once asserted: "The thirteen books of Euclid must have been a tremendous advance, probably even greater than that contained in the *Principia* of Newton."

70° *The most famous single utterance in the history of science.* *The utterance:* If a straight line falling on two straight lines makes the interior angles on the same side together less than two right angles, the two straight lines, if produced indefinitely, meet on that side on which the angles are together less than two right angles.

What is it? It is the *fifth*, or so-called *parallel*, *postulate* of Euclid's *Elements*.

Why is it famous? The lack of terseness and ready comprehensibility of this postulate, when compared with Euclid's other postulates, proved to be the source of much controversy. The dissatisfaction of mathematical scholars with its statement as a postulate is indicated by the fact that many geometers attempted over a period of some twenty centuries either to derive it from Euclid's other postulates and axioms or to replace it by a more acceptable equivalent. This concern over Euclid's fifth postulate furnished the stimulus for the development of a great deal of modern mathematics, and also led to deep and revealing inquiries into the logical and philosophical foundations of the subject. In particular, it led to the liberation of geometry from its traditional mold and to the subsequent creation of many other equally consistent geometries different from that of Euclid. Indeed, it had a similar liberating effect on mathematics as a whole. It dealt a severe blow to the *absolute truth* viewpoint of mathematics and brought forth a new view of the nature of a postulate. Mathematics emerged as an arbitrary creation of the human mind, not as something essentially dictated to us of necessity by the world in which we live.

Who named it so? Cassius J. Keyser, in his *Mathematical Philosophy* (New York: E. P. Dutton and Company, Inc., 1922), p. 113.

71° *The most fruitful single utterance in the history of science.* The question, "Why?"

ARCHIMEDES

ONE of the very greatest mathematicians of all time, and certainly the greatest of antiquity, was Archimedes, a native of the Greek city of Syracuse on the island of Sicily. He was born about 287 B.C. and died during the Roman pillage of Syracuse in 212 B.C. As a youth he studied mathematics at the Museum of Alexandria under the successors of Euclid, after which he returned to his native city to produce his mathematical masterpieces. Some ten treatises by Archimedes have come down to us and there are various traces of lost works. Probably the most remarkable contribution made to mathematics in these works is the early development of some of the methods of the integral calculus.

72° *Archimedes' boast.* It is narrated by Plutarch (ca. A.D. 46–120), in his *Life of Marcellus*, that Archimedes once boasted that if he had a place to stand on he could move the earth. King Hiero of Syracuse was struck with amazement at this statement and asked Archimedes to make good his boast by actually moving some enormous weight. Accordingly a ship, which could not be drawn out of dry dock without the combined labor of a large contingent of men, was loaded with many passengers and a full freight, and Archimedes was asked to move the ship into the water. Seated comfortably in a chair on the beach, the great mathematician and physicist astonished the King and the spectators by effortlessly and singlehandedly causing the ship, by means of a compound pulley arrangement, to slip easily into the sea.

73° *Archimedes' defense of Syracuse.* Impressed with Archimedes' easy success in singlehandedly moving a loaded ship from dry dock to the sea, King Hiero prevailed upon Archimedes to assist in the defense of Syracuse when that city was under the siege directed by the Roman general Marcellus. Many stories have been told of the ingenious contrivances devised by Archimedes to aid his city. There were catapults with adjustable ranges capable of hurling huge projectiles through loopholes in protective bastions. There were movable projecting poles for dropping heavy weights on enemy ships that approached too near the city walls. There were great grappling cranes that hoisted enemy ships from the water and violently shook them till the terrified sailors fell overboard and the ships began to crack up. A story that Archimedes used large burning-glasses to set the enemy's vessels afire is of later origin, but could be true. It has been said that Archimedes had the foes so confounded and frightened that all he had to do was to hang a free end of rope over the city's wall, and the Romans, fearing some horrible contrivance was on the other end, were afraid to go near it. Because of Archimedes' defensive machines, Syracuse held out against the Roman siege and blockade for close to three years. The city's defenses were finally broken only when, during a celebration, the Syracusans overconfidently relaxed their watches.

74° *The fraudulent goldsmith.* Apparently Archimedes was capable of strong mental concentration, and tales are told of his obliviousness to surroundings when engrossed by a problem. Typical is the frequently told story of King Hiero's crown and the suspected goldsmith.

It seems that King Hiero, desiring a crown of gold, gave a certain weight of the metal to a goldsmith, along with instructions. In due time the crown was completed and given to the king. Though the crown was of the proper weight, for some reason the king suspected that the goldsmith had pocketed some of the precious metal and replaced it with silver. The king didn't want to break the crown open to discover if it contained any hidden silver, and so in his perplexity he referred the matter to Archimedes. For a while, even Archimedes was puzzled. Then, one day when in the public baths, Archimedes hit upon the solution by discovering the first law of hydrostatics. In his flush of excitement, forgetting to clothe himself, he rose from his bath and ran home through the streets shouting, "Eureka, eureka" ("I have found it, I have found it").

The famous first law of hydrostatics appeared later as Proposition 7 of the first book of Archimedes' work *On Floating Bodies*. This law, which today every student of physics learns in high school, says that "a body immersed in a fluid is buoyed up by a force equal to the weight of the displaced fluid." This means that of two equal masses of different materials, that one having the greater volume will lose more when the two masses are weighed under water. Thus, since silver is more bulky than gold, it suffers a greater change when weighed under water than does an equal mass of gold. So all Archimedes had to do was to put the crown on one pan of a balance and an equal weight of gold on the other pan, and then immerse the whole in water. In this situation the gold would outweigh the crown if the latter contained any hidden silver. Tradition says that the pan containing the crown rose, and in this way the goldsmith was shown to be dishonest.

75° *The Archimedean screw.* One of Archimedes' inventions, still used in various parts of the world, is the so-called *Archimedean screw*. It consists of a tube (see Figure 8), open at each end, wrapped in

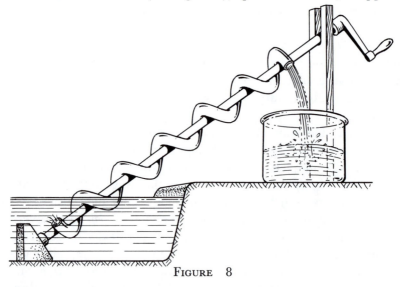

FIGURE 8

helical or cork-screw fashion around a central cylindrical core. One end of the screw is immersed in water and the axis of the screw is inclined to the vertical at a sufficiently large angle. The instrument is then turned, by a handle at its upper end, around its axis. If the inclination of the axis of the screw to the vertical is greater than the pitch of the screw, water will flow along the tube and out the upper end. The Archimedean screw was used in Egypt for irrigating fields and for draining inundated areas. It was also frequently used to empty water from holds of ships.

76° *The stomach of Archimedes.* It has been related that Archimedes worked much of his geometry from figures drawn in the ashes of the hearth and in the after-bathing oil smeared on his body. In connection with the latter, one cannot but wonder how much superb geometry may have been discovered from figures drawn on a human paunch!

77° *The death of Archimedes.* Archimedes met his end during the sack of Syracuse, and several versions of his death have been reported.

According to one account, Archimedes, unaware that Syracuse had been taken by Marcellus, was preoccupied with a diagram drawn on a sand tray when a pillaging Roman soldier entered his apartment. Seeing a shadow cast upon his diagram, Archimedes waved the intruder back and ordered him to stand clear of the figure, whereupon the incensed looter ran a spear through the old man.

Another version says that a Roman soldier entered Archimedes' study and commanded the old man to follow him to Marcellus. Archimedes, in a transport of contemplation upon a problem, declined to do so before he completed his demonstration. The enraged soldier thereupon drew his sword and ran it through the old man.

Still another version reports that a Roman soldier, running upon Archimedes, made to kill the scholar. Archimedes besought the soldier to wait a little, so that he might not leave inconclusive a problem upon which he was engaged. But the soldier, unmoved by the entreaty, instantly killed the great mathematician.

Another account relates that Archimedes was carrying some valuable mathematical instruments to Marcellus when some soldiers saw him and, thinking he carried gold in a vessel, slew the old man.

Plutarch says that nothing afflicted Marcellus so much as the death of Archimedes. The Roman general had given strict orders that no harm was to come to the great man, and one can only imagine the general's ire, and what happened to the GI who disobeyed orders.

78° *The questionable mosaic.*   There is, in the Municipal Art Institute at Frankfurt am Main, a mosaic picturing a venerable scholar seated in a chair before a sand tray resting on a low table. The scholar is looking back over his shoulder at a menacing Roman soldier who, with drawn sword and a pointing arm, seems to be ordering the engaged scholar to rise and follow. There seems little doubt that this mosaic commemorates the last moments of Archimedes.

Until recently, it was believed that the mosaic had been un-covered in the floor of a room of ancient Pompeii (which was destroyed and buried in the volcanic eruption of the year 70), when that city was under excavation not many years ago. This origin of the mosaic is now believed to be fraudulent, and it is thought that the mosaic is a sixteenth-century copy of some earlier work, or perhaps is a pure falsification. Thus another pretty little story has been discredited, or at least cast into grave doubt.

79° *The tomb of Archimedes.*   In his work *On the Sphere and Cylinder*, Archimedes established what are tantamount to our present-day formulas for the volume and the area of a sphere. The basic figure of the work is a sphere of radius r inscribed in a right circular cylinder of radius r and height $2r$ (see Figure 9). Archimedes showed that the volume of the sphere is exactly two thirds of the volume of the circum-scribed cylinder, and the area of the sphere is exactly two thirds of the total area of the circumscribed cylinder. Archimedes was justly proud of these discoveries and expressed a wish that a figure showing a sphere with a circumscribed cylinder (like Figure 9) be engraved on his tombstone. When, with great honor and veneration, Marcellus buried the famous mathematician, he saw to it that Archimedes' request was carried out.

Many years later, in 75 B.C., when Cicero was serving as Roman quaestor in Sicily, he inquired as to the whereabouts of Archimedes' tomb. To his surprise, the Syracusans knew nothing of it and even denied that any such thing existed. With considerable effort and care

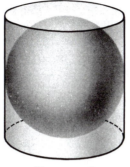

FIGURE 9

Cicero examined all the monuments, of which there were a great many, at the Achradmae gate. Finally he noticed a small column, standing out a little above overgrown briars and shrubs, with the figure of a sphere and circumscribed cylinder upon it. Thus the long-neglected and forgotten tomb of the greatest of all Syracusans was rediscovered. Cicero had men with scythes clear the brush away, and then he left orders that the surrounding grounds be restored and preserved. How long this respect was kept up we do not know; today the tomb has completely vanished.

ERATOSTHENES AND APOLLONIUS

ERATOSTHENES was a native of Cyrene on the south coast of the Mediterranean Sea and was only a few years younger than Archimedes. He spent many years of his early life in Athens and, when about forty, was invited by Ptolemy III of Egypt to come to Alexandria as tutor to his son and to serve as chief librarian at the Museum there. Most of Eratosthenes' mathematical contributions are lost, but we do know of his *sieve* for finding all the prime numbers less than a given number n, his mechanical "mean-finder" for duplicating the cube, and his remarkable measurement of the size of the earth. Some of Archimedes' discoveries were written up in personal letters addressed to Eratosthenes.

Apollonius, who was younger than Archimedes by some twenty-five years, was born about 262 B.C. in Perga in southern Asia Minor.

53

As a young man he went to Alexandria, studied under the successors of Euclid, and remained there for a long time. Later he visited Pergamum in western Asia Minor, where there was a recently founded school and library patterned after that at Alexandria. He returned to Alexandria and died there somewhere around 200 B.C. Although Apollonius was an astronomer of note and although he wrote a variety of mathematical subjects, his chief bid to fame rests on his extraordinary *Conic Sections*, a work which earned him the name, among his contemporaries, of "the Great Geometer."

80° *Eratosthenes' measurement of the earth.* Perhaps the best story that has come down to us about Eratosthenes concerns his determination of the size of the earth. Though there seems to be an element of fortuitousness in Eratosthenes' measurements, the basic idea is simple, ingenious, scientific, and appreciable by students of even the most elementary geometry or trigonometry class.

Eratosthenes learned that at noon, at the summer solstice, a vertical rod held at Syene, a city on the Nile in Egypt, casts no shadow. This fact was confirmed by noting that at that time and place the water in a deep well reflected the sun's rays directly back into the eyes of an observer looking into the well. On the other hand, at noon on the same day, a tall pillar at Alexandria was noted to cast a shadow which showed that the sun's rays there were inclined $\frac{1}{50}$ of a complete circle to the vertical. Now Eratosthenes knew that Syene lay directly south of Alexandria, at a distance of 5000 stadia. The situation we have described is pictured in Figure 10. From a simple proportion we then have

$$\text{circumference of earth: } 5000 = 1 : 1/50.$$

It follows that

$$\text{circumference of earth} = 250,000 \text{ stadia.}$$

There is reason to suppose that a stadium had a length of about one tenth of a mile. This leads to 25,000 miles as the approximate circumference of the earth.

It is interesting that the size of the earth was so simply and so

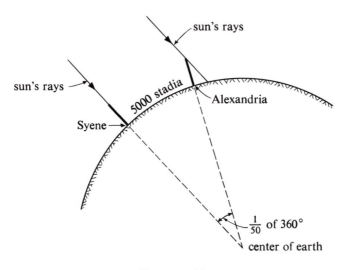

accurately determined without the use of a telescope and without traveling any great distance on land or sea.

81° *The death of Eratosthenes.* It is told that in old age Eratosthenes became either blind or almost blind from opthalmia, and, not wishing to continue life unable to read, he committed suicide by voluntary starvation.

82° *The nicknames of Eratosthenes and Apollonius.* Eratosthenes, who W. W. Rouse Ball has aptly called *the Admirable Crichton* of his age, was singularly gifted in all branches of knowledge. He was distinguished as a mathematician, an astronomer, a geographer, a historian, a philosopher, a poet, and an athlete. It is said that the students at the Museum of Alexandria used to call him *Pentathlus*, the champion in five athletic sports. He was also called *Beta*, and some speculation has been offered as to the possible origin of this nickname. Some believe it was because his broad and brilliant knowledge caused him to be looked upon as a second Plato. A less kind explanation is that, though he was gifted in many fields, he always failed to top his contemporaries in any one branch; in other words, he was always second best. Each of these explanations weakens somewhat when it is learned that a certain astronomer Apollonius, very likely Apollonius of Perga, was called *Epsilon*. Because of this, James Gow has suggested that perhaps Beta and Epsilon arose simply from the Greek numbers (2 and 5) of certain offices or lecture rooms at the Museum particularly associated with the two men. On the other hand, Ptolemy Hephaestio claimed that Apollonius was called Epsilon because he studied the moon, of which the letter ε was a symbol. Of course, the most fitting nickname for Apollonius of Perga is *the Great Geometer*, justly assigned to him by his contemporaries.

83° *The names "ellipse," "parabola," and "hyperbola."* [The following is adapted, with permission, from the article, by Howard Eves, of the same title that appeared in the Historically Speaking section of *The Mathematics Teacher*, April, 1960, pp. 280–281.]

Prior to Apollonius of Perga, the Greeks obtained the conic sections from three types of cones of revolution, according as the vertex angle of the cone was less than, equal to, or greater than a right angle. By cutting each of three such cones by planes perpendicular to a generator of the cone, an ellipse, a parabola, and a hyperbola respectively result. It follows that only one branch of a hyperbola was considered. Apollonius, on the other hand, in Book I of his great treatise *Conic Sections*, obtains all the conic sections in the now familiar way from *one* right or oblique *double* cone.

The names "ellipse," "parabola," and "hyperbola" were supplied by Apollonius, and were borrowed from the early Pythagorean terminology of application of areas. When the Pythagoreans applied a rectangle to a line segment (that is, placed the base of the rectangle along the line segment, with one end of the base coinciding with one end of the segment), they said they had a case of "ellipsis," "parabole," or "hyperbole" according as the base of the applied rectangle fell short of the line segment, exactly coincided with it, or exceeded it. Now let *AB* (see Figure 11) be the principal axis of a conic, *P* any point on the conic, and *Q* the foot of the perpendicular from *P* on *AB*. At *A*, which is a vertex of the conic, draw a perpendicular to *AB* and mark off on it a distance *AR* equal to what we now call the *latus rectum*, or *parameter p*, of the conic (that is, equal to the length of the chord

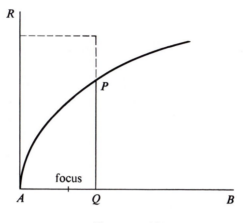

FIGURE 11

which passes through a focus of the conic and is perpendicular to the principal axis of the conic). Apply, to segment AR, a rectangle having AQ for one side and an area equal to $(PQ)^2$. According as the application falls short of, coincides with, or exceeds the segment AR, Apollonius calls the conic an *ellipse*, a *parabola*, or a *hyperbola*. In other words, if we consider the curve referred to a Cartesian coordinate system having its x and y axes along AB and AR respectively and if we designate the coordinates of P by x and y, then the curve is an ellipse if $y^2 < px$, a parabola if $y^2 = px$, and a hyperbola if $y^2 > px$. Actually, in the cases of the ellipse and hyperbola,

$$y^2 = px \mp px^2/d,$$

where d is the length of the diameter through vertex A. Apollonius derives the bulk of the geometry of the conic sections from the geometrical equivalents of these Cartesian equations.

Now an ellipse, being a closed curve lying in the finite part of the plane, has no points in common with the so-called line at infinity in the plane. The parabola, on the other hand, is tangent to the line at infinity and thus has one and only one point in common with that line, and a hyperbola intersects the line at infinity in two distinct points. Because of these relations of the three types of conics with the line at infinity, the adjectives *elliptic*, *parabolic*, and *hyperbolic* have been employed in certain parts of mathematics. Thus, in 1871, Felix Klein called the non-Euclidean geometry of Lobachevsky and Bolyai *hyperbolic geometry*, that of Riemann he called *elliptic geometry*, while the name *parabolic geometry* was reserved for Euclidean geometry. Simplifying historical origins a little, the reason for applying these three epithets to the three geometries is essentially that in the Lobachevsky–Bolyai non-Euclidean geometry there exist two distinct lines through a point P and parallel to a line l not through P, in the Riemann non-Euclidean geometry there are no lines through P parallel to l, in Euclidean geometry there is one and only one line through P parallel to l.

The three adjectives *elliptic*, *parabolic*, and *hyperbolic* are also encountered in projective geometry, and for a similar reason. In projective geometry one studies, among other things, mappings of a line upon itself defined analytically by a symmetrical equation of the form

$$Axx' + B(x + x') + C = 0.$$

Here A, B, C are real constants, and x and x' are the coordinates of corresponding points under the mapping. Such a mapping of a line upon itself is called an *involution*, and of interest in the study of an involution are those points, called *double points*, which map into themselves. To find the double points of the above involution one merely sets $x' = x$, obtaining the quadratic equation

$$Ax^2 + 2Bx + C = 0.$$

The double points of the involution are the real solutions of this quadratic equation. Since the quadratic equation has two distinct real solutions if $B^2 - AC > 0$, one and only one real solution if $B^2 - AC = 0$, and no real solutions if $B^2 - AC < 0$, the involution has come to be called *hyperbolic*, *parabolic*, and *elliptic* in the three cases respectively. Thus a hyperbolic involution has two distinct double points, a parabolic involution has one and only one double point, and an elliptic involution has no double points.

Other uses of the adjective *elliptic* in mathematics occur in the following connections: elliptic cones and cylinders, elliptic coordinates, elliptic functions, elliptic integrals, elliptic paraboloids, elliptic partial differential equations, elliptic points on a surface, and elliptic Riemann surfaces. Similarly we have hyperbolic cylinders, hyperbolic functions, hyperbolic logarithms, hyperbolic paraboloids, hyperbolic partial differential equations, hyperbolic points on a surface, hyperbolic spirals, and hyperbolic Riemann surfaces. And there are parabolic cylinders, parabolic cables, parabolic points on a surface, parabolic spirals, and parabolic Riemann surfaces. The meanings of these mathematical terms can be found, for example, in the *Mathematics Dictionary*, edited by Glenn James and R. C. James (D. Van Nostrand Co., Inc., 1959). In most cases the definitions make clear the·reason for the adoption of the particular adjective employed.

DIOPHANTUS

OF tremendous importance in the history of algebra and of great influence on later European number theorists was Diophantus of Alexandria, a mathematician of uncertain date and nationality. Most historians, upon tenuous evidence, place him in the third century of our era. His most important work is his *Arithmetica*, a treatment of algebraic number theory that marks the author as a genius in this field. This work largely concerns itself with indeterminate algebraic problems where one must find only the rational solutions. Such problems have become known as *Diophantine problems*. Indeed, modern usage of the terminology generally implies the further restriction of the solution to just integers.

84° *Diophantus's personal life.* Beyond the fact that Diophantus flourished at Alexandria, nothing certain is known about his life. There is, however, a problem in the *Greek Anthology* that purports to give us some details. The *Greek*, or *Palatine*, *Anthology* is a collection of forty-six number problems, in epigrammatic form, assembled about 500 A.D. by the grammarian Metrodorus. The problem concerning Diophantus appears as the following epitaphic summary: "Diophantus passed one sixth of his life in childhood, one twelfth in youth, and one seventh more as a bachelor. Five years after his marriage was born a son who died four years before his father, at half his father's age. [How old was Diophantus when he died?]"

If we interpret the phrase "at half his father's age" to mean "at half his father's final age," then, letting x denote Diophantus's age at death, we easily obtain the equation

$$x/6 + x/12 + x/7 + 5 + x/2 + 4 = x,$$

which gives $x = 84$. It then follows that Diophantus spent fourteen years in childhood, seven more in youth, and twelve more as a bachelor, thus marrying when he was thirty-three. Five years later, at the age of thirty-eight, he had a son who died at the age of forty-two when his father was eighty years old. If we should interpret the phrase "at half his father's age" to mean "when he was just half the age of his father," a different solution is obtained with the various episodes occurring at fractional years. Under this interpretation, Diophantus died when $65\frac{1}{3}$ years old and his son when $30\frac{2}{3}$ years old. Out of regard for Diophantus's love of integers, let us allow him and his son the longer spans of life.

85° *The syncopation of algebra.* In 1842, G. H. F. Nesselmann characterized three stages in the historical development of algebraic notation. First we have *rhetorical algebra*, in which the solution of a problem is written, without abbreviations or symbols, as a pure prose argument. Then comes *syncopated algebra*, in which stenographic abbreviations are adopted for some of the more frequently recurring quantities and operations. Finally, as the last stage, we have *symbolic algebra*, in which solutions largely appear in a mathematical shorthand made up of symbols having little apparent connection with the entities they represent. It is fairly accurate to say that algebra prior to the time of Diophantus was rhetorical. One of Diophantus's major contributions to mathematics was the syncopation of Greek algebra. Rhetorical algebra, however, persisted pretty generally in the rest of the world, with the exception of India, for many hundreds of years. Specifically, in western Europe, most algebra remained rhetorical until the fifteenth century. Symbolic algebra made its first appearance in western Europe in the sixteenth century, but did not become prevalent until the middle of the seventeenth century. It is not often realized that much of the symbolism of our elementary algebra textbooks is not more than three hundred years old.

Diophantus had abbreviations for the unknown, powers of the unknown up through the sixth, subtraction, equality, and reciprocals.

Our word "arithmetic" comes from the Greek word *arithmetike*, a compound of the words *arithmos* for "number" and *techne* for "science." It has been rather convincingly pointed out by T. L. Heath that Diophantus's symbol for the unknown was probably derived by merging the first two Greek letters, α and ρ, of the word *arithmos*. This came, in time, to look like the Greek final sigma ς. While there is doubt about this, the meaning of the notation for powers of the unknown is quite clear. Thus "unknown squared" is denoted by Δ^Υ, the first two letters of the Greek word *dunamis* ($\Delta\Upsilon$NAMIΣ) for "power." Again, "unknown cubed" is denoted by K^Υ, the first two letters of the Greek word *kubos* (KYBOΣ) for "cube." Explanations are easily furnished for the succeeding powers of the unknown, $\Delta^\Upsilon\Delta$ (square-square), ΔK^Υ (square-cube), and $K^\Upsilon K$ (cube-cube). Diophantus's symbol for "minus" looks like an inverted V with the angle bisector drawn in. This has been explained as a compound of Λ and I, letters in the Greek word *leipis* (ΛΕΙΨΙΣ) for "lacking." All negative terms in an expression are gathered together and preceded by the minus symbol. Addition is indicated by juxtaposition, and the coefficient of any power of the unknown is represented by the alphabetic Greek numeral following the power symbol. If there is a constant term then M̊, an abbreviation of the Greek word *monades* (MONAΔEΣ) for "units," is used, with the appropriate number coefficient. Thus $x^3 + 13x^2 + 5x$ and $x^3 - 5x^2 + 8x - 1$ would appear as

$$K^\Upsilon \alpha \Delta^\Upsilon \iota \gamma ς \varepsilon \qquad \text{and} \qquad K^\Upsilon \alpha ς \eta \wedge \Delta^\Upsilon \varepsilon \mathring{M} \alpha,$$

which can be read literally as

> unknown cubed 1, unknown squared 13, unknown 5

and

(unknown cubed 1, unknown 8) minus (unknown squared 5, units 1).

It is thus that rhetorical algebra became syncopated algebra.

86° *A diophantine riddle.* The mathematician Augustus De Morgan, who lived in the nineteenth century, when asked the year of his birth, countered with: "I was x years old in the year x^2." When was he born?

87° *The Greek meaning of "arithmetic."* The ancient Greeks made a distinction between the study of the abstract relationships connecting numbers and the practical art of computing with numbers. The former was known as *arithmetic* and the latter as *logistic*. This classification persisted through the Middle Ages until about the close of the fifteenth century, when texts appeared treating both the theoretical and practical aspects of number work under the single name *arithmetic*. It is interesting that today *arithmetic* has its original significance in continental Europe, while in England and America the popular meaning of *arithmetic* is synonymous with that of ancient *logistic*, and in these two countries the descriptive term *number theory* is used to denote the abstract side of number study. In England and America, the word *logistic*, as a noun, refers to symbolic logic, and the word *logistics*, as a noun, is that branch of military art embracing the details of transporting, quartering, and supplying troops in military operations.

THE END OF THE GREEK PERIOD

The immediate successors to Euclid, Archimedes, and Apollonius prolonged the great Greek geometrical tradition for a time, but then it began steadily to languish, and new developments were limited to astronomy, trigonometry, and algebra. Then, toward the end of the third century A.D., 500 years after Apollonius, there lived the enthusiastic and competent Pappus of Alexandria, who strove to rekindle fresh interest in the subject. Although Pappus wrote a number of mathematical commentaries, his really great work is his *Mathematical Collection*, a combined commentary and guidebook of the existing geometrical works of his time, sown with numerous original propositions, improvements, extensions, and historical comments. It is a veritable mine of rich geometrical nuggets and may be called the swan song, or requiem, of Greek geometry, for after Pappus Greek mathematics ceased to be a living study and we find merely its memory perpetuated by minor writers and commentators.

88° *Some famous inequalities.* The Greeks, from early Pythagorean times on, interested themselves in three means, or averages, called the *arithmetic*, the *geometric*, and the *subcontrary*—the last name being later changed to *harmonic* by Archytas and Hippasus. We may define these three means of two positive numbers a and b as

$$A = (a + b)/2, \quad G = \sqrt{ab}, \quad H = 2ab/(a + b),$$

respectively. A favorite question, often asked on a Master's oral examination, is to show that $A \geq G \geq H$, with equality if and only if $a = b$. A singularly neat establishment of these inequalities appears in Book III of Pappus's *Mathematical Collection*. Take B on segment AC (see Figure 12), and erect the perpendicular to AC at B to cut the

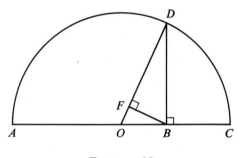

FIGURE 12

semicircle on AC in D. Then, if F is the foot of the perpendicular from B on OD, where O is the midpoint of AC, one may easily show that OD, BD, FD represent the arithmetic mean, the geometric mean, and the harmonic mean of the segments AB and BC. [That OD is the arithmetic mean is obvious. That BD is the geometric mean is well known from high school geometry. That FD is the harmonic mean follows from the similar triangles DFB and DBO, for we have $FD/DB = DB/OD$, whence $FD = (DB)^2/OD = 2(AB)(BC)/(AB + BC)$.] The concerned inequalities are now geometrically evident. The Master's candidate asked to establish these inequalities could hardly do better than to present the Pappus argument, and he would surely make a more impressive showing than by stumbling about with purely algebraic procedures.

89° *Pappus's extension of the Pythagorean Theorem.* [The following is adapted, with permission, from the article, by Howard

Eves, of the same title that appeared in the Historically Speaking section of *The Mathematics Teacher*, November, 1958, pp. 544–546.]

Every student of high school geometry sooner or later becomes familiar with the famous Pythagorean Theorem, which states that *in a right triangle the area of the square described on the hypotenuse is equal to the sum of the areas of the squares described on the two legs.* This theorem appears as Proposition 47 in Book I of Euclid's *Elements*, written about 300 B.C.

Even in Euclid's time, certain generalizations of the Pythagorean Theorem were known. For example, Proposition 31 of Book VI of the *Elements* states: *In a right triangle the area of a figure described on the hypotenuse is equal to the sum of the areas of similar figures similarly described on the two legs.* This generalization merely replaced the three squares on the three sides of the right triangle by any three similar and similarly described figures. A more worthy generalization stems from Propositions 12 and 13 of Book II. A combined and somewhat modernized statement of these two propositions is: *In a triangle, the square of the side opposite an obtuse (acute) angle is equal to the sum of the squares on the other two sides increased (decreased) by twice the product of one of these sides and the projection of the other side on it.* That is, in the notation of Figure 13, $(AB)^2 = (BC)^2 + (CA)^2 \pm 2(BC)(DC)$, the plus or minus sign being taken according as angle C of triangle ABC is obtuse or acute.

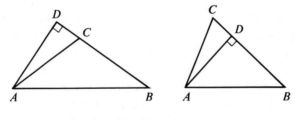

Figure   13

If we employ directed line segments we may combine Propositions 12 and 13 of Book II and Proposition 47 of Book I into the single statement: *If in triangle ABC, D is the foot of the altitude on side BC, then* $(AB)^2 = (BC)^2 + (CA)^2 - 2(BC)(DC)$. Since $DC = CA \cos BCA$, we recognize this last statement as essentially the so-called *law of cosines*, and the law of cosines is indeed a fine generalization of the Pythagorean Theorem.

But perhaps the most remarkable extension of the Pythagorean Theorem that dates back to the days of Greek antiquity is that given by Pappus of Alexandria at the start of Book IV of his *Mathematical Collection*. The Pappus extension of the Pythagorean Theorem is as follows (see Figure 14): *Let ABC be any triangle and CADE, CBFG any*

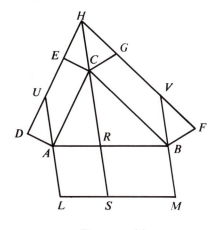

FIGURE 14

parallelograms described externally on sides CA and CB. Let DE and FG meet in H and draw AL and BM equal and parallel to HC. Then the area of parallelogram ABML is equal to the sum of the areas of parallelograms CADE and CBFG. The proof is easy, for we have *CADE = CAUH = SLAR* and *CBFG = CBVH = SMBR.* Hence *CADE + CBFG = SLAR + SMBR = ABML.* It is to be noted that the Pythagorean Theorem has been generalized in two directions, for the right triangle in the Pythagorean Theorem has been replaced by *any* triangle, and the squares on the legs of the right triangle have been replaced by *any* parallelograms.

The student of high school geometry can hardly fail to be interested in the Pappus extension of the Pythagorean Theorem, and the proof of the extension can serve as a nice exercise for the student. Perhaps the more gifted student of geometry might like to try his hand at establishing the further extension (to three-space) of the Pappus extension: *Let ABCD (see Figure 15) be any tetrahedron and let ABD–EFG, BCD–HIJ, CAD–KLM be any three triangular prisms described externally on the faces ABD, BCD, CAD of ABCD. Let Q be the point of intersection of the planes*

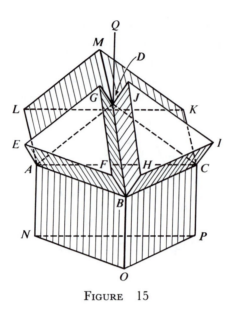

FIGURE 15

EFG, HIJ, KLM, and let ABC–NOP be the triangular prism whose edges AN, BO, CP are translates of the vector QD. Then the volume of ABC–NOP is equal to the sum of the volumes of ABD–EFG, BCD–HIJ, CAD–KLM. A proof analogous to the one given above for the Pappus extension can be supplied.

90° *The first woman mathematician.* Among the minor writers and commentators who succeeded Pappus at the Alexandrian Museum were Theon and his daughter Hypatia. These two lived during the turbulent years of religious upheaval at the end of the fourth and the beginning of the fifth century, and they were the last scholars of mathematics at the famous institution. Theon was the author of a commentary on Ptolemy's *Almagest* and a revision of Euclid's *Elements* from which all our modern editions have derived. His daughter, Hypatia, who undoubtedly obtained an interest in mathematics from her father, is the first woman mathematician to be mentioned in the history of mathematics. She was distinguished in mathematics, medicine, and philosophy, and is reported to have written commentaries on Diophantus's *Arithmetica* and Apollonius's *Conic Sections*.

As an exponent of Alexandrian Neoplatonism, Hypatia lectured on Platonic philosophy at the Museum. Her most notable student was Synesius of Cyrene, who later became bishop of Ptolemais. His affectionate and admiring letters to Hypatia illustrate the close and fruitful contact of early *spiritual* Christianity with pagan philosophy. On the other hand, the bitter and cruel contact of early *militant* Christianity with pagan philosophy is illustrated by Hypatia's death, for in March, 415, she was seized by a mob of infuriated Christians and barbarously murdered on the steps of a pagan temple.

With the tragic passing of Hypatia, the curtain fell on the Greek contribution to mathematics; the long and glorious history of Greek mathematics essentially came to an end.

QUADRANT TWO

*From King Aśoka's columns
to a tantalizing marginal note*

HINDU MATHEMATICS

THE period starting with the fall of the Roman Empire in the middle of the fifth century and extending into the eleventh century is known as Europe's Dark Ages, for during this time civilization in western Europe reached a very low ebb. The period was marked by much physical violence and intense religious faith. Schooling became almost nonexistent and Greek learning all but disappeared. During this period, the history of mathematics, along with that of many other subjects, made a long detour through India and Arabia before it once again took up residence in western Europe.

91° *King Aśoka.* There has appeared, every now and then, a man who definitely was not a mathematician but who nevertheless played an important role in the history of mathematics. Such a man was King Aśoka, an Indian emperor from ca. 274 to ca. 232 B.C. His connection with the history of mathematics lies in the fact that some of his great stone pillars, erected in every important city of India of his day, still stand, and that some of them contain the earliest preserved specimens of our present number symbols that have come down to us.

92° *Inversion.* The Hindus were gifted arithmeticians and made significant contributions to algebra. Many of the arithmetic problems were solved by the method of *inversion*, where one works backward from a given piece of information. Consider, for example, the following problem given during the sixth century by the elder Āryabhata: "Beautiful maiden with beaming eyes, tell me, as thou understandst the right method of inversion, which is the number which multiplied by 3, then increased by $\frac{3}{4}$ of the product, then divided by 7, diminished by $\frac{1}{3}$ of the quotient, multiplied by itself, diminished by 52, by the extraction of the square root, addition of 8, and division by 10 gives the number 2?" By the method of inversion we start with the number 2 and work backward. Thus $[(2)(10) - 8]^2 + 52 = 196$, $\sqrt{196} = 14$, $(14)(\frac{3}{2})(7)(\frac{4}{7})/3 = 28$, the answer. Note that where the problem instructed us to divide by 10 we multiply by 10, where we were told to add 8 we subtract 8, where we were told to extract a square root we take the square, and so forth. It is the replacement of each operation by its inverse that accounts for the name *inversion*. It is, of course, just what we would do if we were to solve the problem by modern methods. Thus, if we let *x* represent the sought number, we have

$$\{\sqrt{[(\tfrac{2}{3})(\tfrac{7}{4})(3x)/7]^2 - 52} + 8\}/10 = 2.$$

To solve this we *multiply* both sides by 10, then *subtract* 8 from each side, then *square* both sides, and so forth.

93° *The rule of three.* The *rule of three*, like much else in elementary arithmetic, seems to have originated with the Hindus, and was actually called by this name by Brahmagupta (ca. 628) and Bhaskara (1114–ca. 1185). For centuries the rule was very highly regarded by merchants. It was mechanically stated without reason, and its connection with proportion was not recognized until the end of the fourteenth century. Here is how Brahmagupta stated the rule: *In the* rule of three, *Argument, Fruit, and Requisition are the names of the terms. The first and last terms must be similar. Requisition multiplied by Fruit, and then divided by Argument, is the Produce.* For clarification consider the following problem given by Bhaskara: If two and a half palas of saffron are purchased for three sevenths of a niska, how many palas will be purchased for nine niskas? Here $\tfrac{3}{7}$ and 9, which are of the same denomination, are the Argument and the Requisition, and $\tfrac{5}{2}$ is the Fruit. The answer, or Produce, is then given by $(9)(\tfrac{5}{2})/(\tfrac{3}{7}) = 52\tfrac{1}{2}$. Today we would regard the problem as a simple application of proportion,

$$x:9 = \tfrac{5}{2}:\tfrac{3}{7}.$$

Much space was devoted to the *rule of three* by early European writers on arithmetic, the mechanical nature of the rule being observable in the doggerel verse and schematic diagrams often used to explain it.

94° *Hindu syncopation of algebra.* The Hindus syncopated their algebra. Like Diophantus, addition was usually indicated by juxtaposition. Subtraction was indicated by placing a dot over the subtrahend, multiplication by writing *bha* (the first syllable of the word *bhavita*, "the product") after the factors, division by writing the divisor beneath the dividend, square root by writing *ka* (from the word *karana*, "irrational") before the quantity. Brahmagupta indicated the unknown by *yā* (from *yāvattāvat*, "so much as"). Unknown integers were prefixed by *rū* (from *rūpa*, "the absolute number"). Additional unknowns were indicated by the initial syllables of words for different colors. Thus a second unknown might be denoted by *kā* (from *kālaka*, "black"), and $8xy + \sqrt{10} - 7$ might appear as

$$yā\ kā\ 8\ bha\ ka\ 10\ rū\ \dot{7}.$$

95° *Bhaskara's daughter*. The Hindu mathematician and astronomer Bhaskara flourished around 1150. Of his works that have come down to us is an arithmetic entitled *Lilavati* ("the beautiful"), and a romantic story is told about this work. According to the tale, the stars foretold dire misfortune if Bhaskara's only daughter Lilavati should marry other than at a certain hour on a certain propitious day. On that day, as the anxious bride was watching the sinking water level of the hour cup, a pearl fell unnoticed from her headdress and, stopping the hole in the cup, arrested the outflow of water. Too late, after the unique lucky moment had passed, the accident was discovered, and the grief-stricken girl was faced with a life of spinsterhood. To console his unhappy daughter, Bhaskara gave her name to his book!

96° *Behold!* Many students of high school geometry have seen Bhaskara's dissection proof of the Pythagorean theorem, in which the square on the hypotenuse is cut up, as indicated in Figure 16, into four triangles each congruent to the given triangle plus a square with side equal to the difference of the legs of the given triangle. The pieces are easily rearranged to give the sum of the squares on the two legs. Bhaskara drew the figure and offered no further explanation than the

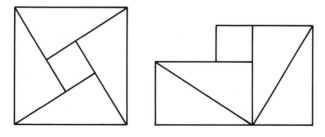

FIGURE 16

73

word "Behold!" Of course, a little algebra supplies a proof. For, if c is the hypotenuse and a and b are the legs of the triangle,

$$c^2 = 4(\tfrac{1}{2}ab) + (b - a)^2 = a^2 + b^2.$$

Perhaps a better "behold" proof of the Pythagorean theorem would be a dynamical one on movie film wherein the square on the hypotenuse is continuously transformed into the sum of the squares on the legs by passing through the stages indicated in Figure 17.

97° *Hindu embroidery.* The Hindus often clothed their arithmetic problems in poetic garb. This was partly because school texts were, for purposes of memorization, written in verse, but particularly because the problems were so frequently propounded as puzzles for sheer social amusement. As a couple of examples, consider the following; the first is adapted from Bhaskara (ca. 1150) and the second from Mahavira(ca. 850).

> The square root of half the number of bees in a swarm has flown out upon a jessamine bush, $\tfrac{8}{9}$ of the swarm has remained behind; one female bee flies about a male that is buzzing within a lotus flower into which he was allured in the night by its sweet odor, but is now imprisoned in it. Tell me, most enchanting lady, the number of bees.

> Into the bright and refreshing outskirts of a forest, which were full of numerous trees with their branches bent down with the weight of flowers and fruits, trees such as jambu trees, lime trees, plantains, areca palms, jack trees, date palms, hintala trees, palmyras, punnago trees, and mango trees—outskirts, the various quarters whereof were filled with many sounds of crowds of parrots and cuckoos found near springs containing lotuses with bees roaming about them—into such forest outskirts a number of weary travelers entered with joy. There were 63 numerically equal heaps of plantain fruits put together and combined with 7 more of those same fruits, and these were equally distributed among 23 travelers so as to have no remainder. You tell me now the numerical measure of a heap of plantains.

Brahmagupta, in connection with his problem offerings, says: "These problems are proposed simply for pleasure; the wise man can invent a thousand others, or he can solve the problems of others by the rules given here. As the sun eclipses the stars by his brilliancy, so the man of knowledge will eclipse the fame of others in assemblies of

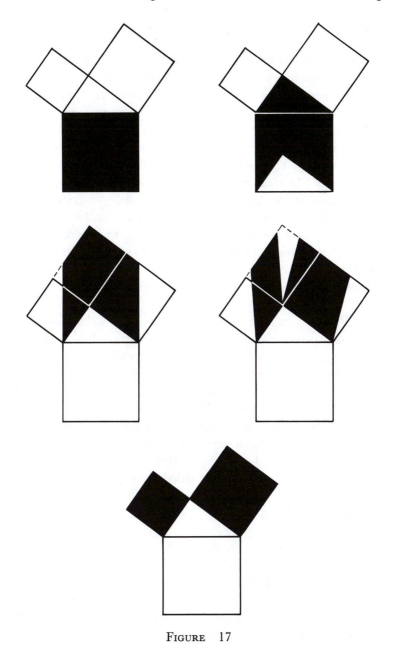

FIGURE 17

75

the people if he proposes algebraic problems, and still more if he solves them.''

Of interest is an early-used *symbolical system of position*, in which numbers were often expressed by suggestive objects. Thus, for 1 might appear the word *moon*, or *Brahma*, or *Creator*, or *form*; for 4, the word *Veda* (because this book was divided into four parts), or *ocean*. As an example, consider the following found in the *Surya Siddhanta*, an anonymous work on astronomy that dates from perhaps the beginning of the fifth century. The number 1,577,917,828 is expressed, from right to left, as: Vasu (a class of eight gods) + two + eight + mountains (the seven mountain ranges) + form + digits (the nine digits) + seven + mountains + lunar days (half of which equals fifteen). Such practice made it possible to represent a number in many different ways, thus greatly facilitating the framing of verses containing arithmetical rules or scientific constants, and making such rules or constants more easy to remember.

98° *Buddha's examination.* Even the very early Hindus exhibited great skill in calculating with large numbers. Thus we are told of an examination to which Buddha, the great religious teacher of about the sixth century B.C., had to submit, in his youth, in order to win the hand of the maiden he loved. He was asked to determine the number of primary atoms which, when placed side by side, would form a line one ''mile'' long. Buddha found the required number as follows: seven primary atoms make a very minute grain of dust, seven of these make a minute grain of dust, seven of these a grain of dust whirled up by the wind, and so on. He proceeded thus, step by step, until he finally reached the length of one ''mile.'' The product of the factors gave a number of fifteen digits.

99° *False position in the Bakshali manuscript.* An anonymous arithmetic, known as the *Bakshali manuscript*, was unearthed in 1881 at Bakshali, in northwest India. It consists of seventy pages of birch bast. Its origin and date have been the subject of much conjecture, estimates of the date ranging from the third to the twelfth century A.D. Some of the problems of the Bakshali manuscript are solved by a reduction to unity, or a sort of *rule of false position*. As an example we find: B gives twice as much as A, C three times as much as B, D four times as much as C; together they give 132; how much did A give? Take 1 for the unknown. Then the amount given by A is 1, that given by B is 2, that by C is 6, that by D is 24, and the sum of these is 33. Divide 132 by 33; the quotient 4 is what A gave.

100° *Contrast between Greek and Hindu mathematics.*   There are many differences between Greek and Hindu mathematics. In the first place, the Hindus who worked in mathematics regarded themselves primarily as astronomers, and thus Hindu mathematics remained largely a handmaiden to astronomy; with the Greeks, mathematics attained an independent existence and was studied for its own sake. Also, due to the caste system, mathematics in India was cultivated almost entirely by the priests; in Greece, mathematics was open to any who cared to study the subject. Again, the Hindus were accomplished computers but mediocre geometers; the Greeks excelled in geometry but, in general, cared little for computational work. Even Hindu trigonometry, which was meritorious, was arithmetical in nature; Greek trigonometry was geometrical in character. The Hindus wrote in verse and often clothed their works in obscure and mystic language; the Greeks strove for clarity and logicality in presentation. Hindu mathematics is largely empirical with proofs or derivations seldom offered; an outstanding characteristic of Greek mathematics is its insistence on rigorous demonstration. Hindu mathematics is of very uneven quality, good and poor mathematics often appearing side by side; the Greeks seemed to have an instinct that led them to distinguish good from poor quality and to preserve the former while abandoning the latter.

A good deal of the contrast between Greek and Hindu mathematics is perpetuated today in the differences between many of our elementary geometry and algebra textbooks. We owe our geometry to the Greeks, our algebra to the Hindus. Only recently, with the emphasis on the "new mathematics," have our algebra textbooks begun to attain a rigor, a logicality, and a qualitative selectiveness of material comparable to that found in the elementary geometry texts.

101° *Srinivasa Ramanujan.* Perhaps the most spectacular Indian mathematician of modern times has been the impoverished clerk and untrained genius Srinivasa Ramanujan (1887–1920), who possessed amazing ability to see quickly and deeply into intricate number relations. He was "discovered" in 1913 by the eminent British number theorist G. H. Hardy (1877–1947), whose efforts brought Ramanujan in the following year to England to study at Cambridge University. There resulted a most remarkable mathematical association between the two men.

The most frequently told story illustrating Ramanujan's uncanny abilities is about a visit once made by Hardy when Ramanujan was ill in a hospital at Putney. Hardy arrived at the hospital in a taxi bearing the seemingly dull number 1729. Hardy took down the number and, in curiosity, asked Ramanujan if there is anything interesting about it. Without hesitation Ramanujan said there certainly is, inasmuch as it is the smallest positive integer that can be represented in two different ways as a sum of two cubes—$1729 = 1^3 + 12^3 = 9^3 + 10^3$.

Ramanujan had the same preternatural ability with numbers that was possessed by his early predecessors, and his work exhibited the same disorganized character, strong intuition, and slighting of deductive processes also found in the earlier men's work. Ramanujan might almost be called a twentieth-century Bhaskara. We see in Ramanujan's work many of the differences between early Hindu and Greek mathematics. Of course, much of this may be traced to the fact that Ramanujan was largely unsystematically self-taught.

ARABIAN MATHEMATICS

Of considerable importance for the preservation of much of world culture was the manner in which the Arabs seized upon Greek and Hindu erudition. The Baghdad caliphs not only governed well, but many became patrons of learning and invited distinguished scholars to their courts. Numerous Hindu and Greek works in astronomy, medicine, and mathematics were industriously translated into the Arabic tongue and thus were saved until later European scholars were able to retranslate them into Latin and other languages. But for the work of the Arabian scholars much of Greek and Hindu science would have been irretrievably lost over the long period of the Dark Ages.

102° *Arabian names in astronomy.* Many names and words used today may be traced back to the Arabian period. Thus anyone interested in observational astronomy probably is aware that a large number of star names, particularly those of the fainter stars, are Arabic. As well-known examples we have Aldebaran, Vega, and Rigel among the brighter stars, and Algol, Alcor, and Mizar among the fainter ones. Many of the star names were originally expressions locating the stars in the constellations. These descriptive expressions, when translated from Ptolemy's catalogue into the Arabic, later degenerated into single words. Thus we have Betelgeuse (armpit of the Central One), Fomalhaut (mouth of the Fish), Deneb (tail of the Bird), Rigel (leg of the Giant), and so forth.

The definitive Greek work on astronomy was written by Claudius Ptolemy of Alexandria about 150 A.D. This very influential treatise, called the *Syntaxis mathematica,* or "Mathematical Collection," was based on the writings of Hipparchus and is noted for its remarkable compactness and elegance. To distinguish it from lesser works on astronomy, later commentators assigned to it the superlative *magiste,* or "greatest." Still later, the Arabian translators prefixed the Arabian article *al,* and the work has ever since been known as the *Almagest.*

103° *The origin of our word "algebra."* Very interesting is the origin of our word *algebra* from the title, *Hisâb al-jabr w'al-muqâbalah,* of al-Khowârizmî's treatise (ca. 825) on the subject. This title has been literally translated as "science of the reunion and the opposition," or more freely as "science of reduction and cancellation." The text, which is extant, became known in Europe through Latin translations, and made the word *al-jabr,* or *algebra,* synonymous with the science of equations. Since the middle of the nineteenth century, *algebra* has come, of course, to mean a great deal more.

The Arabic word *al-jabr,* used in a nonmathematical sense, found its way into Europe through the Moors of Spain. There an *algebrista* was a bonesetter (reuniter of broken bones), and it was usual for a barber of the times to call himself an *algebrista*, for bonesetting and bloodletting were sidelines of the medieval barber.

104° *The origin of our word "algorithm."* In addition to a treatise on algebra, al-Khowârizmî wrote a book on the use of the Hindu numerals, and this second work has also introduced a word into the vocabulary of mathematics. The book is not extant in the original, but in 1857 a Latin translation was found that begins, "Spoken has Algoritmi, ..." Here the name *al-Khowârizmî* has become *Algoritmi*, from which, in turn, was derived our present word "algorism," or "algorithm," meaning the art of calculating in any particular way.

105° *The origin of our word "zero."* Our word *zero* probably comes from the Latinized form *zephirum* of the Arabic *sifr*, which in turn is a translation of the Hindu *sunya*, meaning "void" or "empty." The Arabic *sifr* was introduced into Germany in the thirteenth century by Nemorarius, as *cifra*, from which we have obtained our present word *cipher*.

106° *The origin of our word "sine."* The meanings of the present names of the trigonometric functions, with the exception of *sine*, are clear from their geometrical interpretations when the angle is placed at the center of a circle of unit radius. Thus, in Figure 18, if the radius of the circle is one unit, the measures of tan θ and sec θ are given by the lengths of the *tangent* segment *CD* and the *secant* segment *OD*. And, of course, *cotangent* merely means *c*omplement's tangent, and so on. The functions tangent, cotangent, secant, and cosecant have been known by various other names, these present ones appearing as late as the end of the sixteenth century.

The origin of the word *sine* is curious. Āryabhata called it *ardhā-jyā* ("half chord") and also *jyā-ardhā* ("chord half"), and then abbreviated the term by simply using *jyā* ("chord"). From *jyā* the Arabs phonetically derived *jîba*, which, following the Arabian practice of omitting vowel symbols, was written as *jb*. Now *jîba*, aside from its technical significance, is a meaningless word in Arabic. Later writers, coming across *jb* as an abbreviation for the meaningless *jîba* decided to sub-

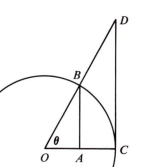

FIGURE 18

stitute *jaib* instead, which contains the same letters and is a good Arabian word meaning "cove" or "bay." Still later, Gherardo of Cremona (ca. 1150), when he made his translations from the Arabic, replaced the Arabic *jaib* by its Latin equivalent, *sinus*, whence came our present word *sine*.

107° *Alhazen's madness.* The name al-Haytham or, more popularly today, Alhazen (ca. 965–1039) has been preserved in mathematics in connection with the so-called *problem of Alhazen*: to draw from two given points in the plane of a given circle lines that intersect on the circle and make equal angles with the circle at that point. The problem leads to a quartic equation that was solved in Greek fashion by an intersecting hyperbola and circle; the problem, like the trisection of a general angle, is beyond the Euclidean tools. Alhazen was born at Basra in South Iraq and was perhaps the greatest of the Moslem physicists. The above problem arose in connection with his *Optics*, a treatise that later had great influence in Europe.

Alhazen unfortunately once boasted that he could construct a machine that would control and regulate the annual inundation of the Nile River. He was accordingly summoned to Egypt by Caliph Hakim to explain and perhaps demonstrate his idea. Aware of the utter impracticality of his scheme, and fearing the anger of the Caliph,

Alhazen feigned madness, for the insane were specially protected in those times. With great care, Alhazen had to keep up the hoax until Hakim's death in 1021.

108° *The three students.* Perhaps the deepest and most original contribution to Arabian algebra was the geometrical solution of cubic equations by Omar Khayyam (ca. 1044–ca. 1123), a native of Khorasan, and known to the Western world through Edward Fitzgerald as the author of the exquisite *Rubaiyat*. Khayyam is also noted for his work on the calendar reform and for a critical treatment of Euclid's *Elements* in which he appears as a forerunner of Saccheri's ideas that finally led to the creation of the first non-Euclidean geometry.

A thought-provoking story has come down to us about Omar Khayyam and two of his schoolmates. In their youth, Nizam ul Mulk, Hasan Ben Sabbah, and Omar Khayyam studied together as pupils of one of the greatest wise men of Khorasan, the Imam Mowaffak of Naishapur. The three youths, all very capable scholars, became close friends. Since it was the belief that a pupil of the Imam stood great chance of attaining fortune, Hasan one day proposed to his friends that the three of them take a vow to the effect that to whomever of them fortune should fall, he would share it equally with the others and reserve no pre-eminence for himself. As the years went by, Nizam proved to be the fortunate one, for he became Vizier to the Sultan Alp Arslan. In time his school friends sought him out and claimed a share of his good fortune according to the school-day vow.

Hasan demanded a governmental post, which was granted by the Sultan at the Vizier's request. But, being selfish and ungrateful, he endeavored to supplant his friend Nizam and was finally disgraced and banished. Omar desired neither title nor office, but simply begged to be permitted to live in the shadow of the Vizier's fortune, where he might promulgate science and mathematics and pray for his friend's long life and prosperity. Impressed by his former schoolmate's modesty and sincerity, the Vizier granted Omar a yearly pension.

After many misadventures and wandering, Hasan became the head of a party of fanatics who, in 1090, seized the castle of Alamut in the mountainous area south of the Caspian Sea. Using the castle as a fortress and center for raids upon passing caravans, Hasan and his band spread terror through the Mohammedan world. Hasan became known as "the old man of the mountain," and it is thought that our present word "assassin" derives either from the leader's name *Hasan* or from the *hashish* opiate with which the band maddened themselves for their murderous assaults. Among the countless victims of the assassins was the old school friend, Nizam ul Mulk.

In contrast to Hasan's turbulent and destructive life, Omar's was tranquil and constructive. He lived peacefully and contributed noteworthily to both the literary and the scientific culture of his age.

So, of the three students, one turned out to be a fine administrator and benefactor, one a miserable renegade and murderer, and one a devoted scholar and creator. Somehow the world at large seems reflected here. If only we could make the odds in favor of the good better than two to one!

109° *Omar's roses.* Omar Khayyam died in Naishapur about 1123. A pupil of his, one Khwajah Nizami of Samarcand, has related that he used to converse with his teacher Omar in a garden, and that Omar once said that his tomb would be located in a spot where the north wind would scatter rose petals over it. Some years later, after the death of his teacher, the former pupil chanced to visit Naishapur and he searched out the master's grave. It was just outside a garden. Boughs of fruit trees hanging over the garden wall had dropped so many blossoms on the grave that the tombstone was completely hidden.

When Edward Fitzgerald, the sympathetic Irish translator who made Omar Khayyam so famous in modern times, passed away in 1883, he was buried in a little English churchyard at Boulge, Suffolk. In 1884, William Simpson, a traveling artist of the *Illustrated London News*, visited Naishapur and found the not-quite-neglected tomb of Omar. Along the edge of the platform in front of the tomb he found some rose-trees and he plucked from these a few of the hips still hanging on the bushes. These seeds, when they arrived in England, were handed over to Mr. Baker of the Kew Botanical Gardens, who planted them and successfully grew some rose-trees from them. On October 7, 1893, one of these rose-trees was transplanted to Fitzgerald's graveside.

Look to the blowing Rose about us—"Lo,
Laughing," she says, "into the world I blow,
    At once the silken tassel of my Purse
Tear, and its Treasure on the Garden throw."

THE RETURN OF MATHEMATICS TO WESTERN EUROPE

Toward the end of the tenth century, Greek classics in science and mathematics began to filter into western Europe. There followed a period of transmission during which the ancient learning preserved by Moslem culture was passed on to the western Europeans. This took place through Latin translations made by Christian scholars traveling to Moslem centers of learning, through the relations between the Norman kingdom of Sicily and the east, and through western European commercial relations with the Levant and the Arabic world.

110° *Gerbert, Pope Sylvester II.* The period starting with the fall of the Roman Empire in the middle of the fifth century and extending into the eleventh century is known as Europe's Dark Ages, for during this period civilization in western Europe reached a very low ebb. Schooling became almost nonexistent, Greek learning all but disappeared, and many of the arts and crafts bequeathed by the ancient world were forgotten. Only the monks of the Christian monasteries, and a few cultured laymen, preserved a slender thread of Greek and Latin learning. The period was marked by much physical violence and intense religious faith. The old order gave way and society became feudal and ecclesiastical.

The Romans had never taken to abstract mathematics, but contented themselves with merely practical aspects of the subject associated with commerce and civil engineering. With the fall of the Roman Empire and the subsequent closing down of much of East–West trade and the abandonment of state engineering projects, even these interests waned, and it is no exaggeration to say that very little in mathematics, beyond the development of the Christian calendar, was accomplished in the West during the whole of the half millennium covered by the Dark Ages.

Of the persons charitably credited with playing a role in the history of mathematics during the Dark Ages, perhaps the greatest was the famous French scholar and churchman Gerbert, who became Pope Sylvester II.

Gerbert was born about 950 in Auvergne, France, and early showed unusual abilities. He was one of the first Christians to study in the Moslem schools of Spain and there is evidence that he may have brought back the Hindu-Arabic numerals, without the zero, to Christian Europe. He possessed great manual talent, and constructed abaci, terrestrial and celestial globes, a clock, and perhaps an organ. Such accomplishments corroborated the suspicions of some of his contemporaries that he had traded his soul to the devil, and fables began to cluster about his name similar to those that later gathered around Faust. There is a story of a speaking statue that foretold Gerbert would die at Jerusalem—a prophecy that was fulfilled (somewhat as in the case of Henry IV of England) by his dying in the Jerusalem church of Rome. In spite of the legends, he steadily rose in the Church and was finally elevated to the papacy in 999. As he successfully steered the Church through the ominous year 1000, he became the first of a long line of so-called magician popes. He died in 1003.

111° *The century of translators.* The loss of Toledo by the Moors to the Christians in 1085 was followed by an influx of Christian scholars to that city to acquire Moslem learning. Other Moorish centers in Spain were infiltrated and the twelfth century became, in the history of mathematics, a century of translators. One of the earliest Christian scholars to engage in this pursuit was the English monk Adelard of Bath (ca. 1120), who studied in Spain and traveled extensively through Greece, Syria, and Egypt. Adelard is credited with Latin translations of Euclid's *Elements* and of al-Khowârizmî's astronomical tables. There are thrilling allusions to the physical risks run by Adelard in his acquisition of Arabic learning; to obtain the jealously guarded knowledge, he disguised himself as a Mohammedan student. Another early translator was the Italian, Plato of Tivoli (ca. 1120), who translated the astronomy of al-Battânî, the *Spherics* of Theodosius, and various other works. The most industrious translator of the period was Gherardo of Cremona (1114-1187), who translated into Latin over ninety Arabian works, among which were Ptolemy's *Almagest*, Euclid's *Elements*, and al-Khowârizmî's algebra. We have already, in Item 106°, mentioned the part played by Gherardo of Cremona in the development of our word *sine*. Other noted translators of the twelfth century were John of Seville and Robert of Chester.

112° *The Norman kingdom of Sicily.* The location and political history of Sicily made that island a natural meeting ground of East and West. Sicily started as a Greek colony, became part of the Roman Empire, linked itself with Constantinople after the fall of Rome, was held by the Arabs for about fifty years in the ninth century, was recaptured by the Greeks, and was then taken over by the Normans. During the Norman regime the Greek, Arabian, and Latin tongues were used side by side, and diplomats frequently traveled to Constantinople and Baghdad. Many Greek and Arabian manuscripts in science and mathematics were obtained and translated into Latin. This work was greatly encouraged by the two rulers and patrons of science, Frederick II (1194–1250) and his son Manfred (ca. 1231–1266).

113° *The Italian commercial centers.* Among the first cities to establish mercantile relations with the Arabic world were the Italian commercial centers of Genoa, Pisa, Venice, Milan, and Florence. Italian merchants came in contact with much of Eastern civilization, picking up useful arithmetical and algebraical information. These merchants played an important part in the dissemination of the Hindu-Arabic numeral system.

114° *From rabbits to sunflowers.* At the threshold of the thirteenth century appeared Leonardo Fibonacci ("Leonardo, son of Bonaccio"), perhaps the most talented mathematician of the Middle Ages. Also known as Leonardo of Pisa (or Leonardo Pisano), Fibonacci was born in the commercial center of Pisa, where his father was connected with the mercantile business. Many of the large Italian businesses in those days maintained warehouses in various parts of the Mediterranean world. It was in this way, when his father was serving as a customs manager, that young Leonardo was brought up in Bougie on the north coast of Africa. The father's occupation early roused in the boy an interest in arithmetic, and subsequent extended trips to Egypt, Sicily, Greece, and Syria brought him in contact with Eastern and Arabic mathematical practices. Thoroughly convinced of the practical superiority of the Hindu-Arabic methods of calculation Fibonacci, in 1202, shortly after his return home, published his famous work called the *Liber abaci*.

The *Liber abaci* is known to us through a second edition that appeared in 1228. The work is devoted to arithmetic and elementary algebra and, though essentially an independent investigation, shows the influence of al-Khowârizmî and Abû Kâmil. The book profusely illustrates and strongly advocates the Hindu-Arabic notation and did much to aid the introduction of these numerals into Europe.

The *Liber abaci* contains a large collection of problems that served later authors as a storehouse for centuries. We have already, in Item 13°, mentioned one interesting problem from the collection, which apparently evolved from a much older problem in the Rhind papyrus. But perhaps the most fruitful problem in the work is the following: "How many pairs of rabbits can be produced from a single pair in a year if every month each pair begets a new pair which from the second month on becomes productive?" Without much effort, the reader can show that this problem leads to the following interesting sequence (wherein the terms are the number of pairs of rabbits present in successive months),

$$1, 1, 2, 3, 5, \ldots, x, y, x + y, \ldots.$$

This sequence, in which the first two terms are 1's and then each succeeding term is the sum of the two immediately preceding ones, has become known as the *Fibonacci sequence* and it has appeared in an astonishing number of unexpected places. It has applications to dissection puzzles, art, and phyllotaxis, and it appears surprisingly in various parts of mathematics.

Consider, for example, the seed head of a sunflower. The seeds are found in small diamond-shaped pockets bounded by spiral curves radiating from the center of the head to the outside edge, as illustrated in Figure 19. Now the curious thing is this, if one should count the number of clockwise spirals and then the number of counterclockwise spirals, these two numbers will be found to be successive terms in the

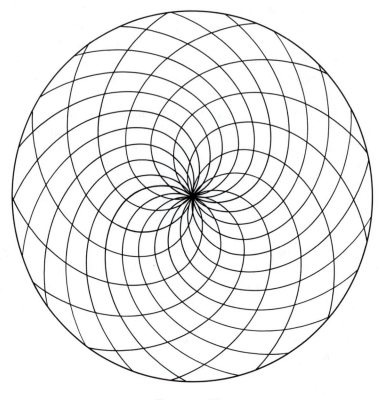

FIGURE   19

Fibonacci sequence. Indeed, this is true of the seed head of any composite flower (for instance a daisy, or an aster); it is more easily tested on a sunflower because these flowers have such large seeds and seed heads. Incidentally, as a further curiosity, the above-mentioned spirals are logarithmic spirals.

Next consider the leaves (or buds, or branches) growing out of the side of a stalk of a plant. If we fix our attention on some leaf near the bottom of the stalk and then count the number of leaves up the stalk until we come to one that is directly over the original leaf, this number is generally a term of the Fibonacci sequence. Also, as we work up the stalk and count the number of times we revolve about the stalk before we come to a leaf directly over the original one, this number is generally the preceding alternate term of the sequence. Similar arrangements occur in a wide variety of plant forms, such as in the leaves of a head of lettuce, in the layers of an onion, and in the conical spirals of a pine cone.

If we form the sequence of ratios of successive terms of the Fibonacci sequence, we obtain

$$\frac{1}{1}, \frac{1}{2}, \frac{2}{3}, \frac{3}{5}, \frac{5}{8}, \frac{8}{13}, \cdots$$

It can be shown mathematically that this sequence of ratios approaches the number

$$r = (\sqrt{5} - 1)/2$$

as a limit. This is the famous so-called *golden ratio* that interested the Greeks of a couple of thousand years ago. The Greeks said that a line segment AC is divided into the golden ratio by the point B if $AB/BC = BC/AC$. In this case it can be shown that each of the ratios AB/BC and BC/AC is equal to r. It seems that nature strives to approximate the golden ratio r.

Psychological tests have shown that to most people the rectangle that appears most pleasing to the eye is the one whose ratio of width to length is the golden ratio r. This rectangle, which may be called the *golden rectangle*, is fundamental in an art technique known as "dynamic symmetry," which has been intensively studied by Jay Hambidge and others. The golden ratio and the golden rectangle have been observed in Greek architecture and Greek pottery, and have been applied to sculpture, painting, architectural design, furniture design, and type display. A number of artists, such as the well known American painter George Bellows, have extensively used the principles of dynamic symmetry in their work.

The Fibonacci sequence finds many unexpected uses in various parts of mathematical study. For instance, there is a computational process, called the *Euclidean algorithm*, for finding the greatest common divisor of two given positive integers. The process requires a number of successive divisions. This number of divisions is relatively small when compared with the magnitudes of the two given positive integers. It is natural to wonder if it is possible to establish a priori a limit for the number of divisions. The answer is given by the following neat theorem due to Gabriel Lamé (1795–1870): *The number of divisions required to find the greatest common divisor of two positive integers is never greater than five times the number of digits in the smaller number.* Now the proof of this theorem utilizes, of all things, some properties of the Fibonacci sequence!

The literature on the Fibonacci sequence and its many properties is incredibly large and continues to grow. The interesting relations seem, like the geometry of the triangle, to be inexhaustible. In fact, in 1963, a group of Fibonacci-sequence enthusiasts, headed by Dr. Verner Hoggatt, Jr., founded the Fibonacci Association and began publication of a journal, *The Fibonacci Quarterly*, devoted principally to research on the Fibonacci sequence. In its first three years of existence, this journal published close to 1000 pages of research in this particular field. In 1968, three extra issues of the journal appeared in a desperate effort to catch up somewhat on the large manuscript backlog.

115° *A mathematical tournament.* Fibonacci's mathematical talents came to the attention of the patron of learning, Emperor Frederick II of the Norman kingdom of Sicily, with the result that Fibonacci was invited to court to partake in a mathematical tournament. Three problems were set by John of Palermo, a member of the emperor's retinue. Fibonacci solved all three problems, a performance that evoked considerable admiration.

The first problem was to find a rational number x such that $x^2 + 5$ and $x^2 - 5$ shall each be squares of rational numbers. Fibonacci gave the answer $x = \frac{41}{12}$, which is correct, since $(\frac{41}{12})^2 + 5 = (\frac{49}{12})^2$ and $(\frac{41}{12})^2 - 5 = (\frac{31}{12})^2$.

The second problem was to find a solution to the cubic equation

$$x^3 + 2x^2 + 10x = 20.$$

Fibonacci attempted a proof that no root of the equation can be expressed by means of irrationalities of the form $\sqrt{(a + \sqrt{b})}$, or, in other words, that no root can be constructed with straightedge and compasses. He then obtained an approximate answer, which, expressed in decimal notation, is

$$1.3688081075,$$

and is correct to nine places. There has been expression of wonder as to how Fibonacci found this answer.

The third problem, which is the easiest of the three, was as follows: "Three men possess a pile of money, their shares being $\frac{1}{2}$, $\frac{1}{3}$, $\frac{1}{6}$. Each man takes some money from the pile until nothing is left. The first man then returns $\frac{1}{2}$ of what he took, the second $\frac{1}{3}$, and the third $\frac{1}{6}$. When the total so returned is divided equally among the men it is found that each then possesses what he is entitled to. How much money was in the original pile, and how much did each man take from the pile?" Here is essentially Fibonacci's solution to the problem. Let s denote the original sum and $3x$ the total sum returned. Before each man received a third of the sum returned, the three men possessed $s/2 - x$, $s/3 - x$, $s/6 - x$. Since these are the sums they possessed after putting back $\frac{1}{2}$, $\frac{1}{3}$, $\frac{1}{6}$ of what they had first taken, the amounts first taken were $2(s/2 - x)$, $\frac{3}{2}(s/3 - x)$, $\frac{6}{5}(s/6 - x)$, and these amounts added together equal s. Therefore $7s = 47x$, and the problem is indeterminate. Fibonacci took $s = 47$ and $x = 7$. Then the sums taken by the men from the original pile are 33, 13, 1.

116° *The blockhead.* Fibonacci sometimes signed his work with *Leonardo Bigollo.* Now *bigollo* has more than one meaning; it means both "traveler" and "blockhead." In signing his work as he did, Fibonacci may have meant that he was a great traveler, for so he was. But a story has circulated that he took pleasure in using this signature because many of his contemporaries considered him a blockhead (for his interest in the new numbers), and it pleased him to show these critics what a blockhead could accomplish.

117° *Finger numbers.* In addition to spoken numbers, *finger numbers* were at one time widely used. Indeed, the expression of numbers by various positions of the fingers and hands probably predates the use of either number symbols or number names. Thus the early written symbols for one, two, three, and four were invariably the suitable number of vertical or horizontal strokes, representing the corresponding number of raised or extended fingers, and the word *digit* (that is, *finger*) for the numbers one through nine can be traced to the same source.

In time finger numbers were extended to include the largest

numbers occurring in commercial transactions, and by the Middle Ages they had become international. In the ultimate development in Europe, the numbers 1, 2, ... , 9 and 10, 20, ... , 90, were represented on the left hand, and the numbers 100, 200, ... , 900 and 1000, 2000, ... , 9000 on the right hand. In this way, any number up to 10,000 was representable by the use of the two hands. Pictures of the finger numbers were given in later arithmetic books. For example, using the left hand, one was represented by partially folding down the little finger, two by partially folding down the little and ring fingers, three by partially folding down the little, ring, and middle fingers, four by folding down the middle and ring fingers, five by folding down the middle finger, six by folding down the ring finger, seven by completely folding down the little finger, eight by completely folding down the little and ring fingers, and nine by completely folding down the little, ring, and middle fingers.

With the above information, the reader can now explain the ninth-century riddle that is sometimes attributed to Alcuin (ca. 775): I saw a man holding eight in his hand, and from the eight he took seven, and six remained. He can also explain the following, found in Juvenal's tenth satire: "Happy is he indeed who has postponed the hour of his death so long and finally numbers his years upon his right hand."

Finger numbers had the advantage of transcending language differences, but, like the vocal numbers, they lacked permanence and were not suitable for performing extended calculations. Nevertheless, there did develop finger processes for certain simple computations. One of these processes, by giving the product of two numbers each between 5 and 10, served to reduce the memory work connected with the multiplication tables. For example, to multiply 7 by 9, raise $7 - 5 = 2$ fingers on one hand and $9 - 5 = 4$ fingers on the other hand. Now add the raised fingers, $2 + 4 = 6$, for the tens digit of the product, and multiply the closed fingers, $3 \times 1 = 3$, for the units digit of the product, giving the result 63. This process is still used by some European peasants, and the reader is invited to prove that the method gives correct results.

118° *The eulogist of mathematics.* Roger Bacon (ca. 1214–ca. 1294), original genius that he was, had little ability in mathematics but was acquainted with many of the Greek works in geometry and astronomy, and, as his eulogies attest, fully appreciated the value of the subject. It was in his *Opus Majus* that he made what has become his best known panegyric on mathematics: "Mathematics is the gate and key of the sciences.... Neglect of mathematics works injury to all knowledge, since he who is ignorant of it cannot know the other sciences or the things of this world. And what is worse, men who are thus ignorant are unable to perceive their own ignorance and so do not seek a remedy."

119° *Submathematical analysis*. Although European mathematics during the Middle Ages was essentially practical, speculative mathematics did not entirely die out. The meditations of scholastic philosophers led to subtle theorizing on motion, the infinitely large and the infinitely small, the continuous and the discrete, all of which are fundamental concepts in modern mathematics. The centuries of scholastic disputes and quibblings may, to some extent, account for the remarkable transformation from ancient to modern mathematical thinking, and might, as suggested by E. T. Bell, constitute a *submathematical analysis*. From this point of view, Thomas Aquinas (1226–1274), perhaps possessing the most acute mind of the thirteenth century, can well be considered as having played a part in the development of mathematics.

THE FOURTEENTH, FIFTEENTH, AND SIXTEENTH CENTURIES

THE fourteenth century was a mathematically barren one. It was the century of the Black Death, which swept away more than a third of the population of Europe, and in this century the Hundred Years' War, with its political and economical upheavals in northern Europe, got well under way.

The fifteenth century witnessed the start of the European Renaissance in art and learning. With the collapse of the Byzantine Empire, culminating in the fall of Constantinople to the Turks in 1453, refugees flowed into Italy bringing with them treasures of Greek civilization. Many Greek classics, hitherto known only through the often inadequate

Arabic translations, could now be studied from original sources. Also, about the middle of the century, printing was invented and revolutionized the book trade, enabling knowledge to be disseminated at an unprecedented rate. Toward the end of the century, America was discovered and soon the earth was circumnavigated. Mathematical activity was largely centered in the Italian cities and in the central European cities of Nuremberg, Vienna, and Prague, under the influence of trade, navigation, astronomy, and surveying.

In the sixteenth century the development of arithmetic and algebra continued, the most spectacular mathematical achievement of the century being the discovery, by Italian mathematicians, of the algebraic solution of cubic and quartic equations. Algebra began to pass from the syncopated stage into the symbolic stage.

120° *The mechanical eagle.*   Some men eminent in the history of mathematics achieved additional fame as inventors or constructors of various mechanical devices. One recalls Archimedes (ca. 287–212 B.C.) and his screw pump and devices to defend Syracuse against the Romans. And there was Heron of Alexandria (75?), that encyclopedic writer on mathematical and physical subjects, who designed some hundred machines and toys, such as a siphon, a steam engine, a fire engine that pumped water, an altar fire that was mechanically lighted as soon as the temple doors were opened, a wind organ, a robot that poured wine, and mirrors possessing all sorts of bizarre properties. There was also Gerbert (ca. 950–1003), who became Pope Sylvester II and who constructed abaci, terrestrial and celestial globes, a clock, and perhaps an organ. Simon Stevin (1548–1620) astonished the citizens of his day with a sailwagon. John Napier (1550–1617) was the science fiction man of his age and previsioned, at least on paper, the modern machine gun, war tank, and submarine. Galileo (1564–1642) made telescopes, invented the first modern type microscope, and designed the once very popular sector compasses. William Oughtred (1574–1660) invented circular and straight slide rules. Blaise Pascal (1623–1662) constructed the world's first adding machines and invented the one-wheeled wheelbarrow. The great Dutch genius Christiaan Huygens (1629–1695) made clocks and watches. The youthful Isaac Newton (1642–1727) was remarkably inventive, making—among other things—

kites that carried lanterns, little grist mills run by mice, and toys for his friends. Gottfried Wilhelm Leibniz (1646–1716) invented a calculating machine that multiplied.

Another man of the same sort was Johann Müller (1436–1476), the ablest and most influential mathematician of the fifteen century, and more generally known, from the Latinized form of his birthplace of Königsberg ("king's mountain"), as Regiomontanus. At a young age he studied under Peuerbach in Vienna and was later entrusted with the task of completing the latter's translation of the *Almagest*. He also translated, from the Greek, works of Apollonius, Heron, and Archimedes. His treatise *De triangulis omnimodis*, written about 1464 but posthumously published in 1533, is his greatest mathematical achievement and was the first systematic European exposition of plane and spherical trigonometry considered independently of astronomy.

Regiomontanus traveled much in Italy and Germany, finally settling in 1471 at Nuremberg. It was there that he built an astronomical observatory and established a printing press. But the device that won him most admiration was a mechanical eagle that flapped its wings and saluted Emperor Maximilian I when that monarch entered Nuremberg. This mechanical eagle, which exhibited considerable mechanical ingenuity, was regarded as one of the marvels of the age.

In 1475, Regiomontanus was invited to Rome by Pope Sixtus IV to partake in the reformation of the calendar. Shortly after his arrival, at the age of 40, he suddenly died. Some mystery shrouds his death, for, though most accounts claim he probably died of a pestilence, it was rumored that he was poisoned by an enemy.

121° *Introduction of + and −*. The first appearance in print of our present + and − signs was in an arithmetic, published in Leipzig in 1489, by Johannes Widman (born ca. 1460 in Bohemia). Here the signs are not used as symbols of operation but merely to indicate excess and deficiency. These signs were used in the same way in some slightly earlier written manuscripts that Widman had studied.

It may be that the plus sign is a contraction of the Latin word *et*, which was frequently used to indicate addition, and it may be that the minus sign is contracted from the abbreviation $\bar{m}$ for minus. Again, the minus sign may be the simple hyphen that was used by merchants to separate the indication of the weight of the receptacle from the total weight of the merchandise. Other more or less plausible explanations have been offered.

The + and − signs were used as symbols of algebraic operation in 1514 by the Dutch mathematician Van der Hoeke but were probably so used earlier.

122° *The cossic art.*   Many mathematicians of the fifteenth and sixteenth centuries, following the practice of Fibonacci and the Arabs, called the unknown quantity the *thing*—in Italian, *cosa*. It is for this reason that algebra was sometimes designated as the *cossic art*. For example, in 1525, Christoff Rudolff wrote an algebra entitled *Die Coss*. This algebra was very influential in Germany and an improved edition of it was brought out by Michael Stifel (1486–1567) in 1553. It was in Rudolff's book that our familiar radical sign (adopted perhaps because it resembles a small *r*, for *radix*) was introduced.

123° *Leonardo da Vinci's proof of the Pythagorean Theorem.*   There is a clever proof of the Pythagorean Theorem claimed to have been devised by the great artist Leonardo da Vinci (1452–1519). It is a congruency-by-subtraction proof, based on the diagram of Figure 20. Since, in that figure, the quadrilaterals *IFGH*, *IFBA*, *CJDA*, and *JCBE* are congruent, it follows that the hexagons *ABFGHI* and *ACBEJD* have equal areas. But hexagon *ABFGHI* is composed of the squares on the legs of the right triangle *ABC* along with two triangles congruent to triangle *ABC*; and hexagon *ACBEJD* is composed of the square on the hypotenuse of the right triangle *ABC* along with two triangles congruent to triangle *ABC*. It now follows that the sum of the squares on the legs of the right triangle *ABC* is equal to the square on the hypotenuse.

There is a work, *De diuina proportione*, of the Italian friar Luca Pacioli (ca. 1445–ca. 1509) that contains figures of the regular solids thought to have been drawn by Leonardo da Vinci. The book was published in 1509. Seldom is an author of a mathematical text able to secure such an outstanding artist to illustrate his work.

124° *The stone upon which one may sharpen his wits.*   The most influential British textbook writer of the sixteenth century was

Robert Recorde (ca. 1510–1558). Recorde wrote in English, his works appearing as dialogues between master and student. He studied at Oxford and then took a medical degree at Cambridge. He taught mathematics in private classes at both institutions while in residence there, and after leaving Cambridge he served as physician to Edward VI and Queen Mary.

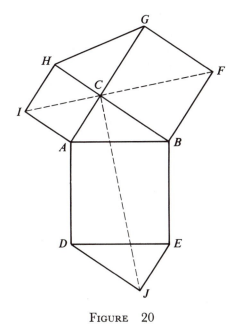

FIGURE 20

Recorde wrote at least four mathematical texts, each bearing a finely suggestive title: an arithmetic entitled *The Grovnd of Artes* (1542), which enjoyed at least 29 printings; an astronomy named *The Castle of Knowledge* (1551), which was one of the first works to introduce the Copernican system to English readers; a geometry called *The Pathewaie to Knowledge* (1551), which contained an abridgment of Euclid's *Elements*; and an algebra with the especially fine title of *The Whetstone of Witte* (1557), in which our modern symbol for equality was used for the first time.

On the title page of *The Whetstone of Witte* are two stanzas of verse, among which are the lines:

Here if you lift your wittes to whette,
Moche sharpenesse therby shall you gette.
Dulle wittes hereby doe greately mende,
Sharpe wittes are fined to their fulle ende.
Now proue, and praise, as you doe finde,
And to your self be not vnkinde.

125° *The origin of our equal sign.* Recorde justified his adoption of a pair of equal parallel line segments for the symbol of equality "bicause noe 2 thynges can be moare equalle." Of further interest here is a remark made by Recorde in his geometry work. There he says, "Parallels, or gemowe [i.e. twin] lynes be suche lines as be drawn foorth still in one distaunce, and are no nerer in one place then in an other, for if they be nerer at one ende then in the other, then are they no parallels." Thus the equality of the spaces between the two left ends and between the two right ends of the line segments of his equality symbol may also have influenced his choice of the symbol. Later, when the inequality symbols (> and <) were introduced by Thomas Harriot (1560–1621), that writer may have chosen his symbols because of the inequality of the spaces between the two left ends and between the two right ends of the involved line segments. (See Item 149°.)

126° *The death of Robert Recorde.* In later life Robert Recorde became "Comptroller of the Mines and Monies" in Ireland. His last years were spent in poor health in prison, and though for many years it was supposed he was imprisoned for personal debt, it now seems that he may have committed some misdemeanor in connection with his work in Ireland. Apparently he foresaw his end, for his algebra text, *The Whetstone of Witte*, closes with the lines:

> *Master.* But harke, what meaneth that hastie knockyng at the doore?
> *Scholar.* It is a messenger.
> *Master.* What is the message; tel me in mine eare. Yea Sir is that the matter. Then is there noe remedie, but that I must neglect all studies and teaching, for to withstande those daungers. My fortune is not so good, to haue quiete tyme to teache.... I mighte haue been quietly permitted, to rest but a little lōger.

Robert Recorde died in the King's Bench Prison in 1558.

127° *Adam Riese.* With the interest in education that accompanied the Renaissance and with the tremendous increase in commercial activity at the time, hosts of popular textbooks in arithmetic began to appear. At least three hundred such books were printed in Europe prior to the seventeenth century. These texts were largely of two types, those written in Latin by classical scholars often attached to the Church schools, and those written in the vernaculars by practical teachers interested in preparing boys for commercial careers. These latter teachers often also served as town surveyors, notaries, and gaugers, and included the influential Rechenmeisters supported by the Hanseatic League, a powerful protective union of commercial towns in the Teutonic countries.

The greatest of all the Rechenmeisters of the sixteenth century was Adam Riese (1492–1559). As the most influential of the German writers of arithmetics, he was in Germany what Robert Recorde was in England. His commercial arithmetic published in 1522 ran through at least 37 editions before 1600. So reputable was this work that even today in Germany the phrase *nach Adam Riese* is used to indicate arithmetical skill and accuracy.

A cute geometrical story is told about Adam Riese. It seems that one day Riese and a draftsman entered into a friendly contest to see which one of them could, with straightedge and compasses, draw more right angles in one minute. The draftsman drew a straight line and then proceeded, by the standard method now taught in high school, to erect perpendiculars to the line. Adam Riese, on the other hand, drew a semicircle on a straight line and then in rapid order drew a large number of inscribed right angles. Riese easily won the contest.

128° *Nicolaus Copernicus.* There are a number of stories concerning efforts of a student on behalf of his teacher. Such a story is told of Georg Joachim Rhaeticus (1514–1576) and his former teacher Nicolaus Copernicus (1473–1543). Rhaeticus had once studied for two years under Copernicus and became an enthusiastic supporter of his teacher's heliocentric theory of the universe. If it hadn't been for the energetic importunities of Rhaeticus, Copernicus would never have seen his great work in print. It is said that the first copy off the press was rushed to Copernicus as he lay on his deathbed and that just before he lapsed into insensibility the copy was placed in the dying author's hands.

129° *Michael Stifel.* Michael Stifel (1486–1567) has been described as the greatest German algebraist of the sixteenth century. His best known mathematical work is his *Arithmetica integra*, published in 1544. In the first part of this work, Stifel points out the advantages of associating an arithmetical progression with a geometrical one, thus foreshadowing the invention of logarithms by John Napier nearly a century later.

Stifel was one of the oddest personalities in the history of mathematics. He was originally a monk, was converted by Martin Luther, and became a fanatical reformer. His erratic mind led him to indulge in number mysticism. From an analysis of Biblical writings, he prophesied the end of the world on October 3, 1533. He convinced a large number of believing peasants to abandon their work and property and to wait with him on the appointed day on a neighboring hilltop, where a chariot was to touch down and conduct him and his followers to heaven. Feeling a growing lack of confidence as the day progressed, Stifel excused himself from his worried believers for a moment, raced into town, and coaxed the local constabulary there to lock him safely in jail. His life was thus saved from the later bitter anger of the peasants whose lives he had ruined.

From this we learn that if you *must* predict the end of the world, be sure to select a date well beyond the end of your possible life span. This sage advice was followed in later years by John Napier (1550–1617), who claimed the Creator proposed to end the world some time in the years between 1688 and 1700.

130° *The art of beasting.* An extreme example of Stifel's mystical reasoning is his proof, by arithmology, that Pope Leo X was the "beast" mentioned in the *Book of Revelations*: "Let him that hath understanding count the number of the beast: for it is the number of a man; and his number is six hundred three score and six."

From LEO DECIMVS, Stifel retained L, D, C, I, M, V, since these letters have significance in the Roman numeral system. He then added X, for Leo X and because *Leo decimvs* contains ten letters, and omitted the M, because it stands for *mysterium*. A rearrangement of the letters gave DCLXVI, or 666, the "number of the beast" in the *Book of Revelations*. This discovery gave Stifel such extreme comfort that he believed his interpretation must have resulted from an inspiration from God.

Some years later, Napier, the inventor of logarithms, showed that 666 stands for the Pope at Rome, and his Jesuit contemporary, Father Bongus, declared that it stands for Martin Luther. Father Bongus's reasoning ran as follows. If from A to I represents 1 to 9, from K to S represents 10 to 90 (by tens), and T to Z represents 100 to 500 (by hundreds),* we obtain

M	A	R	T	I	N		L	V	T	E	R	A
30	1	80	100	9	40		20	200	100	5	80	1

which gives, as a sum, 666.

During World War I, arithmology was used to show that 666 must be interpreted as Kaiser Wilhelm. It has been shown that 666 spells Nero when expressed in the letter symbols of the Aramaic language in which the *Book of Revelations* was originally written.

Augustus De Morgan said that when it comes to the use of interpretations of one branch of the Church against another, the true explanation of the three sixes is that the interpreters are "six of one and half a dozen of the other."

THE EPISODE OF CUBIC AND QUARTIC EQUATIONS

SOME cubic equations are found on ancient Babylonian tablets, and Archimedes cleverly solved a cubic equation that arose in his work *On the Sphere and Cylinder*. Also, Omar Khayyam solved cubic equations geometrically insofar as the positive roots are concerned. But it was the Italian algebraists of the sixteenth century who first solved general cubic and quartic equations in terms of the coefficients of the equations. The story of this discovery, when told in its most colorful version, rivals any pages ever written by Benvenuto Cellini.

---

* The Latin alphabet is like the English, except that it lacks j and w. Moreover, in the upper-case letters, a U appears as a V.

131° *The story of the algebraic solution of cubic equations.* Briefly told, the facts seem to be these. About 1515, Scipione del Ferro (1465–1526), a professor of mathematics at the University of Bologna, solved algebraically the cubic equation $x^3 + mx = n$, probably basing his work on earlier Arabic sources. He did not publish his result but revealed the secret to his pupil Antonio Fior. Now about 1535, Nicolo of Brescia, commonly referred to as Tartaglia ("the stammerer") because of a childhood injury which affected his speech, claimed to have discovered an algebraic solution of the cubic equation $x^3 + px^2 = n$. Believing this claim was a bluff, Fior challenged Tartaglia to a public contest of solving cubic equations, whereupon the latter exerted himself and only a few days before the contest found an algebraic solution for cubics lacking a quadratic term. Entering the contest equipped to solve two types of cubic equations, whereas Fior could solve but one type, Tartaglia triumphed completely. Later Girolamo Cardano, an unprincipled genius who taught mathematics and practiced medicine in Milan, upon giving a solemn pledge of secrecy, wheedled the key to the cubic from Tartaglia. In 1545, Cardano published his *Ars magna*, a great Latin treatise on algebra, at Nuremberg, Germany, and in it appeared Tartaglia's solution of the cubic. Tartaglia's vehement protests were met by Lodovico Ferrari, Cardano's most capable pupil, who argued that Cardano had received his information from del Ferro through a third party and accused Tartaglia of plagiarism from the same source. There ensued an acrimonious dispute from which Tartaglia was perhaps lucky to escape alive.

Since some of the actors in the above drama seem not always to have had the highest regard for truth, one finds a number of variations in the details of the plot.

132° *Girolamo Cardano.* Girolamo Cardano is one of the most extraordinary characters in the history of mathematics. He was born in Pavia in 1501 as the illegitimate son of a jurist and developed into a man of passionate contrasts. He commenced his turbulent professional life as a doctor, studying, teaching, and writing mathematics while practicing his profession. He once traveled as far as Scotland and upon his return to Italy successively held important chairs at the Universities of Pavia and Bologna. He was imprisoned for a time for heresy because he published a horoscope of Christ's life, showing that whatever Christ did he had to by the dictation of his stars. Resigning his chair in Bologna he moved to Rome and became a distinguished astrologer, receiving, oddly enough, a pension as astrologer to the papal court. He died in Rome in 1576, by drinking poison, one story says, so as to fulfill his earlier astrological prediction of the date of his death. Many stories are told of his wickedness, as when in a fit of rage he cut off the ears of his younger son. At about the same time, his older son was executed for murder. An inveterate gambler, Cardano wrote a gambler's manual in which are considered some interesting questions on probability. It may be that history has somewhat maligned Cardano. Cardano's auto-biography, of course, supports this view.

133° *Tartaglia.*   Tartaglia had a hard childhood. He was born about 1499 at Brescia to very poor parents. He was present at the taking of Brescia in 1512 by the French. After the capture of the town, most of the inhabitants sought sanctuary in the cathedral, but were there massacred by the soldiers. Tartaglia's father, who was a postal messenger of the town, was among those who were killed. The boy suffered some severe saber cuts that split his skull in three places and cleft his jaws and palate. He was left for dead, but when his mother, who had hidden elsewhere, got to the cathedral to search out her family, she found him still alive and managed to carry him off. Lacking all resources she recalled that a dog when wounded always licks the injured place. So for days she licked the poor boy's head. He ultimately recovered, but the injury to his palate left him with an impediment in his speech, and it was from this that he received his nickname of Tartaglia, the stammerer.

It was only with great sacrifice that the boy was able to educate himself. His mother scraped together sufficient money to send him to school for fifteen days, along with the advice that he must make the best of his opportunity. This he interpreted by stealing a copybook from which he later taught himself how to read and write. It is said that lacking the means to buy paper, he was obliged to use the tombstones in the cemetery as slates.

Tartaglia later earned his livelihood teaching science and mathe-matics in various Italian cities. He was a gifted mathematician. In addition to his success in solving cubic equations, he is credited with being the first to apply mathematics to the science of artillery fire. He wrote what is generally considered the best Italian arithmetic of the sixteenth century, and he published editions of Euclid and Archimedes. He died in Venice in 1557.

134° *The story of the algebraic solution of quartic equations.* It was not long after the cubic had been solved that an algebraic solution was discovered for the general quartic equation. In 1540, the Italian mathematician Zuanne da Coi proposed the following problem to Cardano: "Divide 10 into three parts such that they shall be in continued proportion and that the product of the first two shall be 6." If the three parts be denoted by a, b, c, we have

$$a + b + c = 10, \quad ac = b^2, \quad ab = 6.$$

If a and c are eliminated we obtain the quartic equation

$$b^4 + 6b^2 + 36 = 60b.$$

Although Cardano was unable to solve the equation, his pupil Ferrari succeeded, and Cardano had the pleasure of publishing this solution, as well as Tartaglia's solution of the cubic, in his *Ars magna*.

FRANÇOIS VIÈTE

THE greatest French mathematician of the sixteenth century was François Viète, frequently called by his semi-Latin name of Vieta, a lawyer and member of parliament who devoted most of his leisure time to mathematics. He wrote a number of works on trigonometry, algebra, and geometry, most of which were printed and distributed at his own expense. He was born in 1540 at Fontenay and died in 1603 in Paris.

135° *The origin of a friendship.* Some entertaining anecdotes are told about Viète. Thus there is the story about the ambassador from the Low Countries who boasted to King Henry IV that France had no mathematician capable of solving a problem proposed in 1593 by his countryman Adrianus Romanus (1561–1615) and which required the solution of a 45th degree equation. Viète was summoned and shown the equation. Recognizing an underlying trigonometric connection he was able, in a few minutes, to give two roots, and then later gave twenty-one more. The negative roots escaped him. In return Viète challenged Romanus to solve the problem of Apollonius (to draw a circle tangent to three given circles), but Romanus was unable to obtain a solution using Euclidean tools. When he was shown the proposer's elegant solution he traveled to Fontenay to meet Viète with the result that a warm friendship developed.

136° *Christian versus unchristian.*   There is also the story of how Viète successfully deciphered a Spanish code containing several hundred characters and for two years France profited thereby in its war with Spain. So certain was King Philip II that the code was undecipherable that he complained to the Pope that the French were employing magic against his country, "contrary to the practice of the Christian faith." We note that there was no complaint that the war was contrary to the practice of the Christian faith.

137° *Work unfit for a Christian.*   In his *De numerosa potestatum resolutione* of 1600, Viète gives a systematic process for successively approximating to a root of an algebraic equation. Though the method was in general use until about 1680, the procedure becomes so laborious for equations of high degree that one seventeenth-century mathematician described it as "work unfit for a Christian."

SIMON STEVIN, JOHN NAPIER, AND HENRY BRIGGS

Many of the fields in which numerical calculations are important, such as astronomy, navigation, trade, engineering, and war, have made ever increasing demands that these computations be performed more quickly and accurately. These increasing demands were met successively by four remarkable inventions: the Hindu-Arabic notation, decimal fractions, logarithms, and the modern computing machines. The earliest systematic treatment of the second of these devices was given by Simon Stevin in his influential text on arithmetic of 1585 entitled *La Disme*. The third of the great labor-saving devices was the invention of John Napier and appeared in his brochure of 1615 entitled *Mirifici logarithmorum canonis descriptio* (A Description of the Wonderful Law of Logarithms). Henry Briggs helped to perfect Napier's invention and devoted great energies toward the construction of the first table of common logarithms.

Simon Stevin was born at Bruges in 1548. In later life he served as quartermaster general of the Dutch Army and directed many public works. He was the most influential mathematician of the Low Countries in the sixteenth century and he won a high reputation for his work on statics and hydrostatics. He died at the Hague in 1620.

John Napier, who was born in 1550 when his father was only sixteen years of age, lived most of his life at the imposing family estate of Merchiston Castle, near Edinburgh, Scotland, and expended most of his energies in the political and religious controversies of his day. He was violently anti-Catholic and championed the causes of John Knox and James I. As relaxation from his political and religious polemics, Napier amused himself with the study of mathematics and science. He died in 1617.

Henry Briggs was born near Halifax, England, in 1561. He had the honor of being the first occupant of a chair of geometry founded by Sir Thomas Gresham in 1596 at Gresham College in London and the first occupant of a chair of geometry founded by Sir Henry Savile in 1619 at Oxford University. Briggs died at Oxford in 1631.

138° *A multiple reputation.* In the history of mathematics, Stevin is best known as one of the earliest expositors of the theory of decimal fractions. In the history of physics he is best known for his contributions to statics and hydrostatics. To the savants of his time he was best known for his works on fortifications and military engineering. To the general populace of his time he was best known for his invention of a carriage propelled by sails, which ran along the seashore carrying 28 people and easily outstripping a galloping horse.

139° *Napier's misjudgment of himself.* In 1593, Napier published a bitter and widely read attack on the Church of Rome entitled *A Plaine Discouery of the whole Reuelation of Saint Iohn*, in which he endeavored to prove that the Pope was Anti-Christ and that the Creator proposed to end the world in the years between 1688 and 1700. The book ran through twenty-one editions, at least ten of them during the author's lifetime. Napier sincerely believed that his reputation with posterity would rest upon this book. How wrong he proved to be! His book is today totally disregarded and known only to a curious few. Instead, his reputation today rests solidly, widely, and almost solely upon one of his mathematical diversions, the invention of logarithms.

140° *The science fiction writer of his day.* Napier wrote prophetically of various infernal war engines, accompanying his writings with plans and diagrams. He predicted the future would develop a piece of artillery that could "clear a field of four miles circumference of all living creatures exceeding a foot of height," that it would produce "devices for sayling under water," and that it would create a chariot with "a living mouth of mettle" that would "scatter destruction on all sides." In World War I these were realized as the machine gun, the submarine, and the army tank, respectively.

141° *Exposing a thief.* It is no wonder that Napier's remarkable ingenuity and imagination led some to believe he was mentally unbalanced and others to regard him as a dealer in the black art. Many stories, probably unfounded, are told in support of these views. Thus there was the time he announced that his coal black rooster would identify to him which one of his servants was stealing from him. He put his rooster in a box in a darkened room and instructed the servants to enter one by one and to place a hand on the rooster's back. Napier assured his servants that his rooster would expose the culprit to him at the completion of these performances. Now, unknown to the servants, Napier had coated the bird's back with lampblack. The innocent servants, having nothing to fear, did as they were bidden, but the guilty one decided to protect himself by not touching the bird. In this way he was exposed, for he was the only servant to return from the darkened room with clean hands.

142° *Impounding pigeons.* Napier was annoyed by a neighbor's pigeons that flew onto his land and ate his grain. He accordingly asked the neighbor to restrict the pigeons' flights, otherwise he would have to impound the birds as payment for the stolen grain. The neighbor, believing the capture of his pigeons to be virtually impossible, said that Napier was welcome to the birds if he could catch them. Imagine the neighbor's surprise when next day he observed his pigeons staggering and reeling on Napier's lawn and Napier calmly going about collecting them and putting them into a large sack. Napier had soaked some peas in wine or brandy and scattered them about his lawn. He then found it easy to collect the drunk birds and put them in the sack.

143° *The meeting.* Briggs so admired Napier's invention of logarithms that he decided to travel from London to Edinburgh to meet the ingenious Scotsman. Briggs was delayed on the journey and the awaiting Napier complained to a common friend, "Ah, John, Mr. Briggs will not come." At that very moment a knock was heard at the gate and Briggs was ushered into Napier's presence. For almost a quarter of an hour each man beheld the other without speaking a word. Then Briggs said, "My lord, I have undertaken this long journey purposely to see your person, and to know by what engine of wit or ingenuity you came first to think of this most excellent help in astronomy, namely, the logarithms, but, my lord, being by you found out, I wonder nobody found it out before, when now known it is so easy." Briggs remained at Merchiston Castle for a month as Napier's guest.

144° *Some terminology.* The word *logarithm* means "ratio number," and was adopted by Napier after first using the expression *artificial number.* Briggs introduced the word *mantissa* (for the decimal part of a logarithm), which is a late Latin term of Etruscan origin, originally meaning an "addition" or "makeweight," and which in the sixteenth century came to mean "appendix." The term *characteristic* (for the integral part of a logarithm) was also suggested by Briggs and used by Adriaen Vlacq (1600–1666), a Dutch bookseller and publisher who helped complete the table of logarithms started by Briggs. It is curious that it was customary in early tables of common logarithms to print the characteristic as well as the mantissa, and that it was not until the eighteenth century that the present custom of printing only the mantissa was established.

145° *Laplace's statement.* Napier's wonderful invention of logarithms was enthusiastically adopted throughout Europe. In astronomy, in particular, the time was overripe for such a discovery. As Pierre-Simon Laplace (1749–1827) once asserted, the invention of logarithms "by shortening the labors doubled the life of the astronomer."

146° *A historical curiosity.* Nowadays a logarithm is universally regarded as an exponent. Thus if $n = b^x$, we say x is the logarithm of n to the base b. From this definition the laws of logarithms follow immediately from the laws of exponents. One of the curiosities of the history of mathematics is the fact that logarithms were discovered before exponents were in use.

147° *Napierian logarithms versus natural logarithms.* As we know today, the power of logarithms as a computing device lies in the fact that by them multiplication and division are reduced to the simpler operations of addition and subtraction. A forerunner of this idea is apparent in the formula

$$\sin A \sin B = \tfrac{1}{2}[\cos (A - B) - \cos (A + B)],$$

well known in Napier's time, and it is quite probable that Napier's line of thought started with this formula, since otherwise it is difficult to account for his initial restriction of logarithms to those of the sines of angles. Napier labored at least twenty years upon his theory, and, whatever the genesis of his idea, his final definition of a logarithm is as follows. Consider a line segment AB and an infinite ray DE, as shown in Figure 21. Let points C and F start moving simultaneously from A and D, respectively, along these lines, with the same initial rate. Suppose C moves with a velocity always numerically equal to the

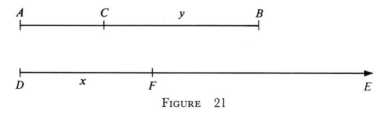

FIGURE   21

distance CB, and that F moves with a uniform velocity. Then Napier defined DF to be the logarithm of CB. That is, setting $DF = x$ and $CB = y$,

$$x = \text{Nap log } y.$$

In order to avoid the nuisance of fractions, Napier took the length of AB as 10^7, for the best tables of sines available to him extended to seven places. From Napier's definition, and through the use of knowledge not available to Napier, it develops that*

$$\text{Nap log } y = 10^7 \log_{1/e} (y/10^7),$$

so that the frequently made statement that Napierian logarithms are natural logarithms is actually without basis. One observes that the Napierian logarithm decreases as the number increases, contrary to what happens with natural logarithms.

It further develops that, over a succession of equal periods of time, y *decreases* in *geometric* progression while x *increases* in *arithmetic* progression. Thus we have the fundamental principle of a system of logarithms, the association of a geometric and an arithmetic progression. It now follows, for example, that if $a/b = c/d$, then

$$\text{Nap log } a - \text{Nap log } b = \text{Nap log } c - \text{Nap log } d,$$

which is one of the many results established by Napier.

It was when Briggs visited Napier that both men agreed that the tables would be more useful if they were altered so that the logarithm of 1 would be 0 and the logarithm of 10 would be an appropriate

---

* The result is easily shown with the aid of a little calculus. Thus we have $AC = 10^7 - y$, whence velocity of $C = -dy/dt = y$. That is, $dy/y = -dt$, or integrating, $\ln y = -t + C$. Evaluating the constant of integration by substituting $t = 0$, we find $C = \ln 10^7$, whence

$$\ln y = -t + \ln 10^7.$$

Now

$$\text{velocity of } F = dx/dt = 10^7,$$

so that $x = 10^7 t$. Therefore

$$\text{Nap log } y = x = 10^7 t = 10^7 (\ln 10^7 - \ln y)$$
$$= 10^7 \ln (10^7/y) = 10^7 \log_{1/e} (y/10^7).$$

power of 10. Thus were born the so-called *Briggsian*, or *common*, logarithms of today. Logarithms of this sort, which are essentially logarithms to the base 10, owe their superior utility in numerical computations to the fact that our number system also is based on 10. For a number system having some other base *b* it would, of course, be most convenient for computational purposes to have tables of logarithms also to the base *b*.

THOMAS HARRIOT AND WILLIAM OUGHTRED

THOMAS Harriot (1560–1621) is usually considered the founder of the English school of algebraists. His great work in this field, the *Artis analyticae praxis*, was not published until ten years after his death. This work did much toward setting the present standards for a textbook on the theory of equations. William Oughtred (1574–1660) was one of the most influential of the seventeenth-century English writers on mathematics. In the same year, 1631, that Harriot's work on algebra appeared, there also appeared the first edition of Oughtred's popular *Clavis mathematicae*, a work on arithmetic and algebra that did much toward spreading mathematical knowledge in England. Oughtred placed emphasis on mathematical symbols, giving over 150 of them, of which only three have come down to present times: the cross (×) for multiplication, the four dots (::) used in a proportion, and our frequently used symbol for "difference between" (~).

148° *Harriot in America.* Thomas Harriot is of special interest to Americans because in 1585 he was sent by Sir Walter Raleigh as a surveyor with Sir Richard Greenville's expedition to the New World to map what was then called Virginia but is now North Carolina.

149° *On the origin of > and <.* During his stay of roughly a year in America, Harriot took the opportunity to study the Indians and to learn to speak their language. Upon returning to England, he wrote a book entitled *A Brief and True Report of the New Found Land of Virginia, of the Commodities, and of the Nature and Manner of the Naturall Inhabitants* (first edition 1588, second edition 1590). Captain John

White, who accompanied Harriot, made sketches of scenes and people seen by the two men. In the 1590 edition of Harriot's book appeared engravings made by Thomas de Bry of some of the sketches drawn by White. One of these engravings shows a rear view of an Indian chief on whose left shoulder blade appears the mark reproduced in Figure 22. If the small serif-like marks are removed, and the resulting symbol

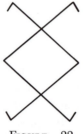

FIGURE   22

pulled apart in the horizontal direction, there will appear two symbols similar to those that Harriot chose for "is greater than" and "is less than." It is thus possible, as was pointed out by Charles L. Smith of the State University of New York at Potsdam, that a mark on the back of an Indian chief suggested to Harriot two mathematical symbols that have now been in use for more than three centuries.

In the absence of any stated or recorded motivation on the part of Harriot, the above explanation could well be the true one. But there is at least one feature of Harriot's early symbols perhaps militating against the conjecture. Harriot constructed his inequality signs as very long, horizontally drawn-out symbols, and not at all like the short stubby symbols appearing on the Indian chief's back.

Of course, since Harriot had adopted the long drawn-out equality sign of Robert Recorde, it could be that his long drawn-out inequality signs were so designed merely for similitude of representation. However, one would like to think that Harriot had a more rational motivation for the origin of his symbols than an adaptation of marks appearing on the back of an Indian chief. Such a rational and easily conceived motivation would be this: In an expression like 2 = 2, the space between the two left ends of the bars of the equality sign is equal to the space between the two right ends of these bars, and, also, the number on the left of the equality sign is equal to the number on the right. Therefore, in designing a symbol to represent the qualitative relation between 4 and 2, say, since the left number 4 is greater than the right number 2, why not adopt a symbol composed of two *converging* bars, so that the space between the two left ends of these bars is greater than that between the two right ends of the bars? Because of Harriot's adoption of the long equality sign, a long inequality sign for "is greater than," composed of two converging bars like we have just described, should, to circumvent possible misinterpretation, *completely* converge, yielding, over the years, 4 > 2.

Whatever Harriot's motivation might have been for the origin of his inequality signs, the motivation described immediately above has fine pedagogical value, and once a student hears this motivation he will never confound the two symbols > and < .

150° *The teacher of giants.* Oughtred was, by profession, an Episcopal minister. He is reputed to have been a "pittiful preacher," but his sermons rapidly improved when the gathering Puritan Revolution shrank his congregation and threatened his livelihood. He preferred to work at mathematics and gave free private lessons to pupils interested in the subject. Among his pupils were John Wallis, Christopher Wren, and Seth Ward, later famous, respectively, as a mathematician, an architect, and an astronomer.

151° *The invention of the slide rule.* In his work *The Circles of Proportion* (1632), Oughtred described a circular slide rule. He was not, however, the first to describe in print a slide rule of the circular type, and an argument of priority of invention rests between him and Richard Delamain, one of his pupils. But Oughtred does seem unquestionably to have invented, about 1622, the straight logarithmic slide rule. In 1620, Edmund Gunter (1581–1626) constructed a logarithmic scale, or a line of numbers on which the distances are proportional to the logarithms of the numbers indicated, and mechanically performed multiplications and divisions by adding and subtracting segments of this scale with the aid of a pair of dividers. The idea of carrying out these additions and subtractions by having two like logarithmic scales, one sliding along the other as shown in Figure

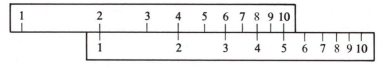

FIGURE   23

23, is due to Oughtred. Although Oughtred invented such a simple slide rule as early as 1622, he did not describe it in print until 1632. A runner for the slide rule was suggested by Isaac Newton in 1675, but was not actually constructed until nearly a century later. Several slide rules for special purposes, such as for commercial transactions, for measuring timber, and so forth, were devised in the seventeenth century. The log log scale was invented in 1815, and it was in 1850 that the French army officer Amédée Mannheim (1831–1906) standardized the modern sliding rule.

152° *Oughtred's longevity.* In speaking of his college days, Oughtred said:

> The time which over and above those usuall studies I employed upon the Mathematical sciences, I redeemed night by night from my naturall sleep, defrauding my body, and inuring it to watching, cold, and labour, while most others tooke their rest.

And Aubrey, in his *Brief Lives*, gives this description of Oughtred:

> He was a little man, had black haire, and blacke eies (with a great deal of spirit). His head was always working. He would drawe lines and diagrams in the dust . . . did use to lye a bed till eleaven or twelve a clock . . . Studyed late at night; went not to bed till 11 a clock; had his tinder box by him; and on the top of his bed-staffe, he had his inke-horne fix't. He slept but little. Sometimes he went not to bed in two or three nights.

Oughtred thus seems to have ignored the usual rules of good health, and probably continued to ignore them throughout his long life. When he finally died, it is said that he did so in a transport of joy at receiving the news of the restoration of Charles II. To this Augustus De Morgan once remarked, "It should be added, by way of excuse, that he was eighty-six years old."

GALILEO GALILEI AND JOHANNES KEPLER

THERE were two outstanding astronomers who contributed notably to mathematics in the early part of the seventeenth century, the Italian, Galileo Galilei (1564–1642), and the German, Johannes Kepler (1571–1630). To Galileo we owe the modern spirit of science as a harmony between experiment and theory. He founded the mechanics of freely falling bodies and laid the foundation of dynamics in general, a foundation upon which Isaac Newton was able later to build the science. To Kepler we owe one of the most remarkable inductions in science, his famous three laws of planetary motion, arrived at after extraordinary pertinacity and after much computational labor, trial and error, and false solutions. These laws of planetary motion are landmarks in the history of astronomy and mathematics, for in an effort to justify them Isaac Newton was led to create modern celestial mechanics.

153° *The oscillating lamp.* At the age of seventeen, Galileo was sent by his parents to the University of Pisa to study medicine. One day, while attending a service in the cathedral at Pisa, his mind was distracted by a great bronze lamp which was suspended from the ceiling and which oscillated to and fro with changing amplitude. Using the beat of his pulse to keep time, he was surprised to find that the period of an oscillation of the lamp was independent of the size of the arc of oscillation. By experiments, he later showed that the period of a swinging pendulum is also independent of the weight of the pendulum's bob, and thus depends solely on the length of the pendulum. It is said that Galileo's interest in science and mathematics was roused by this problem, and then further stimulated by the chance attendance at a lecture on mathematics at the University. The result was that he asked for, and secured, parental permission to abandon medicine and to devote himself to science and mathematics instead, fields in which he possessed strong natural talent.

154° *Falling bodies.* When Galileo was twenty-five, he received an appointment as professor of mathematics at the University of Pisa. It was while holding this appointment that he made his famous public experiments with falling bodies. Before a crowd of students, faculty, and priests, he dropped two pieces of metal, one ten times the weight of the other, from the top of the leaning tower of Pisa. The two pieces of metal struck the ground at practically the same moment, thus contradicting Aristotle, who said that a heavier body falls faster than a lighter one. But even the visual evidence of Galileo's experiment did not shake the faith of the other professors at the University in the teaching of Aristotle. The authorities at the University were so shocked at Galileo's sacrilegious insolence in contradicting Aristotle that they made life unpleasant for him there, with the result that in 1591 he resigned his professorship. The following year he accepted a professorship at the University of Padua, where there was an atmosphere more friendly to scientific pursuits. Here, for nearly eighteen years, Galileo continued his experiments and his teaching, and won widespread fame.

155° *The telescope, and further trouble.*   About 1607, an apprentice to the spectacle maker, Hans Lippershay of Holland, while playing with some of his master's spectacle lenses, discovered that if he held two of the lenses at an appropriate distance apart, objects seen through the pair of lenses became enlarged. The apprentice brought his discovery to the attention of his master, who placed two lenses in a tube and displayed the device as a toy in his shop window. The toy was seen by a government official, who bought it and presented it to Prince Maurice of Nassau. As commander of the armed forces of the United Netherlands, Prince Maurice saw the possibilities of the toy as a spyglass for military use.

By 1609, news of the invention of the spyglass reached Galileo, who soon made a spyglass greatly superior to the one made by Lippershay. Upon request, he demonstrated his instrument in Venice, where, from the top of the highest church in the city, Venetian senators were able to see the sails of an approaching ship a full two hours before they were visible by naked eye. Galileo presented his model to the Doge of Venice, who, like Prince Maurice, recognized the immense possibilities of the instrument in naval and military operations, and Galileo was given a sizably increased stipend.

Galileo went on and made four more telescopes, as his instruments were named (from the Greek *tele*, "far," *skopos*, "watching"), each more powerful than the last. With the fifth telescope, which had a power of thirty diameters, Galileo noticed, on the night of January 7, 1610, two small stars to the east of the planet Jupiter and one to the west. The following night, to his surprise, all three stars were to the west of the planet, and three nights later he found there was still another small star revolving about Jupiter. He had discovered Jupiter's four bright satellites and was furnished a striking confirmation of the Copernican theory of smaller bodies revolving about larger ones. But this discovery only aroused once more the bigoted opposition of many churchmen, who accepted the authority of Aristotle; Aristotle had asserted that the earth, and hence man, is the center of the universe. One churchman even accused Galileo of placing the four satellites of Jupiter inside his telescope!

When Galileo named the four satellites "the Medicean stars," after the family name of the ruler of Tuscany, the Grand Duke of Tuscany was flattered and offered Galileo a munificent sinecure. Galileo accepted the offer and unwisely left the relatively free atmosphere of Venetia for the orthodox closeness of Tuscany.

At Florence, with the aid of his telescope, Galileo made further discoveries that confirmed the Copernican theory. He also discovered the existence of sunspots, which again contradicted Aristotle who claimed the sun is spotless and without blemish. Trouble was again brewing for Galileo. Copernicus's book was placed on the Index of prohibited works, there to remain for two hundred years, and Galileo was advised to cease upholding the Copernican theory. Nevertheless, in 1632, Galileo published his findings and his confirmation of the Copernican theory in his famous work which has come to be briefly referred to as the *Discorsi*.

156° *The inquisition and the unhappy end of a great scholar.* Not all churchmen were against Galileo and his discoveries. There were some more enlightened men, like Pope Gregory XV and Cardinal Barberini, who later became Pope, who did not oppose his views. In fact, Cardinal Barberini actually confirmed some of Galileo's discoveries by observing them through the telescope. Nevertheless, the forces of reaction gained the upper hand. By 1632, when Galileo published his famous *Discorsi*, in which the Copernican theory appears to triumph over the Ptolemaic theory, Cardinal Barberini had become Pope Urban VIII and now paid heed to the reactionaries who suggested that the character Simplicius ("the simpleton") in Galileo's book represented him as supporter of the Aristotelian-held Ptolemaic theory of the universe. The sale of Galileo's book was prohibited and a commission was appointed to look into the matter. The report was unfavorable and condemned Galileo for "maintaining that the earth moves and that the sun is stationary." Galileo was arrested and summoned before the Inquisition. On June 22, 1633, an ill and an old man, he was forced, under threat of torture, to declare, "I adjure, curse, and detest the said errors and heresies and generally every error and sect contrary to the said Holy Church; and I swear that I will nevermore in future say or assent anything verbally or in writing which may give rise to a similar suspicion of me; but that if I know any heretic, or any one suspected of heresy, I will denounce him to this Holy Office or to the Inquisitor and Ordinary of the place in which I may be."

Having thus perjured his conscience, the old scholar's life was broken. He was permitted to continue innocuous scientific work, but became blind and died in January, 1642, still under the supervision of the Inquisition and a virtual prisoner in his own home.

There is a legend that, as Galileo rose to his feet after his forced recantation and denial of the earth's motion, he muttered softly under his breath to himself, "The earth *does* move all the same." Whatever the basis of this story, it has come to be a sort of proverb to the effect that truth shall prevail despite all attempts at suppression. And so it came to pass, for the year 1642, which saw the death of Galileo in captivity, saw also the birth of Isaac Newton.

157° *Authority versus reasoning in science.* Galileo has been quoted as saying: "In questions of sciences, the authority of a thousand is not worth the humble reasoning of a single individual."

158° *Galileo's reconciliation of science and Scripture.* All his life long, Galileo was a religious man and a devout Catholic. It accordingly distressed him to find the views to which he was irresistibly led by his observations and reasonings as a scientist condemned as contradicting the Scriptures of the Church of which he considered himself a loyal member. He therefore felt impelled to think out for himself the relation between science and Scripture. Many Christian scientists have, from time to time, found themselves in this position. It occurred, for example, in the middle of the nineteenth century when difficulties were felt in reconciling Darwin's theory of evolution with the Biblical account of the creation of living things.

Galileo's conclusion was that the Bible is not, and never was intended to be, a textbook on astronomy, or biology, or any other science. In short, it was not intended as a book to teach us scientific truths that we can discover for ourselves. Rather, it was intended as a book to reveal spiritual truths that we could not have found out for ourselves. Now the conflict between science and Scripture lies in the fact that these spiritual truths are expressed in the Bible in ways natural to the people to whom, and through whom, they were originally revealed. But this is clearly just an accident of time and should therefore be overlooked. A scientist should not be upset to find the Bible picturing the world in a way natural to the early Hebrews, and a churchman should not be upset to find a scientist picturing the world in a way contrary to the description in the Bible. The way in which the world is described is entirely incidental to the real aim of the Bible, and no way is inconsistent with the spiritual teachings of the Bible.

159° *Some Galileo–Kepler correspondence.* When Kepler's work *Mysterium cosmographicum*, which openly advocated the Copernican theory, was published in 1596, the author sent a copy to Galileo. On August 4, 1597, Galileo wrote to Kepler thanking him for the book:

> I would certainly dare to approach the public with my ways of thinking if there were more people of your mind. As this is not the case, I shall refrain from doing so . . . .
>
> Yours in sincere friendship,
> GALILACUS GALILAEUS
> *Mathematician at the Academy of Padua*

A couple of months later, on October 13, 1597, Kepler replied to Galileo urging his fellow Copernican to be bold and proceed openly with his beliefs:

Be of good cheer, Galileo, and appear in public. If I am not mistaken, there are only a few among the distinguished mathematicians of Europe who would disassociate themselves from us . . . .

But, as we have seen in Item 156°, Galileo chose a later day for his day of reckoning.

160° *Tycho Brahe's golden nose.* In 1600 Kepler became associated with the eminent Danish-Swedish astronomer Tycho Brahe (1546–1601) as that man's assistant. Brahe was a hard man to work for as he was a vitriolic fellow of violent temper.

Some people have a glass eye, or a wooden leg; Brahe had a golden nose. Earlier in life, when he was located at the University of Rostock, a quarrel he had with a Danish nobleman led to a duel in which Brahe had the misfortune to lose a sizable piece of his nose. Brahe had the lost portion replaced by a piece of material composed of wax, silver, and gold.

When Brahe suddenly died in October, 1601, Kepler inherited both his master's position and his vast and very accurate collection of astronomical data on the motion of the planets.

161° *Kepler's pertinacity.* It has often been said that almost any problem can be solved if one but continuously worries over it and works at it a sufficiently long time. Somewhat as Thomas Edison said of invention being one percent inspiration and ninety-nine percent perspiration, problem solving is one percent imagination and ninety-nine percent perseverance. Perhaps nowhere in the history of science is this more clearly demonstrated than in Kepler's incredible pertinacity in solving the problem of the motion of the planets about the sun. Thoroughly convinced of the Copernican theory that the planets revolve in orbits about the central sun, Kepler strenuously sought to determine the nature and position of those orbits, and the manner in which the planets travel in their orbits. After many highly imaginative attempts (see, for example, Item 64°), made when he had little data to aid in verification, Kepler inherited Tycho Brahe's enormous mass of very accurate observations of the motion of the planets. The problem then became this: to obtain a pattern of motion of the planets that would exactly jibe with Brahe's great set of observations. So dependable were Brahe's recordings that any solution that should differ from Brahe's observed positions by even so little as a quarter of the moon's apparent diameter must be discarded as incorrect. Kepler had, then, first to guess with his *imagination* some plausible solution, and then with painful *perseverance* to endure the mountains of tedious calculation needed to confirm or reject his guess. He made hundreds of fruitless attempts and performed reams and reams of calculations, laboring with undiminished zeal and patience for twenty-two years. Finally he solved his problem, in the form of his three famous laws of planetary motion:

I. *The planets move about the sun in elliptical orbits with the sun at one focus.*

II. *The radius vector joining a planet to the sun sweeps over equal areas in equal intervals of time.*

III. *The square of the time of one complete revolution of a planet about its orbit is proportional to the cube of the orbit's semimajor axis.*

The empirical discovery of these laws from Brahe's mass of data constitutes one of the most remarkable inductions ever made in science. With justifiable pride, Kepler prefaced his *Harmony of the Worlds* of 1619 with the following poetic outburst:

I am writing a book for my contemporaries or—it does not matter—for posterity. It may be that my book will wait for a hundred years for a reader. Has not God waited for 6000 years for an observer?

162° *The rarity of problem solvers.* One now sees why there are so few expert problem solvers. An expert problem solver must be endowed with two incompatible qualities, a restless imagination and a patient pertinacity.

163° *Pure versus applied mathematics.* One never knows when a piece of pure mathematics may receive an unexpected application. As William Whewell once said, "If the Greeks had not cultivated the conic sections, Kepler could not have superseded Ptolemy." It is very interesting that 1800 years after the Greeks had developed the properties of the conics merely to satisfy their intellectual cravings, there should occur such an illuminating practical application of them.

164° *A life of misfortune.* It is sad that Kepler's personal life was made almost unendurable by a multiplication of worldly misfortunes. An infection with smallpox when he was but four years old left his eyesight much impaired. In addition to his general lifelong weakness, he spent a joyless youth, his marriage was a constant source of unhappiness, his favorite child died of smallpox, his wife went mad and died, he was expelled from his lectureship at the University of Grätz when that city fell to the Catholics, his mother was charged and imprisoned for witchcraft and for almost a year he desperately tried to save her from the torture chamber, he himself very narrowly escaped condemnation of heterodoxy, and his stipend was always in arrears. One report says that his second marriage was even less fortunate than his first although he took the precaution to analyze carefully the merits and demerits of eleven girls before choosing the wrong one. He was forced to augment his income by casting horoscopes, and he died of a fever while on a journey to obtain some of his long overdue salary.

165° *Numerology and theology.* It seems that in one respect numerology and theology are alike: it does not necessarily make any difference to a man's science what he believes or disbelieves about either. Some of the leading twentieth-century scientific numerologists are as distinguished in science as are their opponents who have only disrespect for all number mysticism.

GÉRARD DESARGUES AND BLAISE PASCAL

GÉRARD Desargues (1593–1662) and Blaise Pascal (1623–1662) were forerunners of the great nineteenth-century researchers in the field of projective geometry. Desargues was an engineer, architect, and one-time French army officer who, when he was in his thirties and living in Paris, made a considerable impression on his contemporaries through a series of gratuitous lectures on geometry that later led to a book. Among those who appreciated his work was the youthful Blaise Pascal, who was to become an eminent French mathematician, physicist, and man of letters. In fact, Pascal once credited Desargues as being the source of much of his inspiration. Desargues and Pascal died in the same year; Desargues was 69, but Pascal was only 39.

166° *Desargues' forgotten book.* In 1639, nine years after Kepler's death, there appeared in Paris a remarkably original but little heeded treatise on the conic sections, written by Gérard Desargues. The work was so generally neglected by other mathematicians that it was soon forgotten and all copies of the publication disappeared. Two centuries later, when the French geometer Michel Chasles wrote his history of geometry, there was no means of estimating the value of Desargues' work. Six years later, however, in 1845, Chasles happened upon a manuscript copy of the treatise, made by Desargues' pupil, Philippe de la Hire, and since that time the work has been recognized as a classic in the early development of synthetic projective geometry.

Several reasons can be advanced to account for the initial neglect of Desargues' little volume. It was overshadowed by the more supple analytic geometry introduced by Descartes two years earlier. Geometers were generally expending their energies either developing this new powerful tool or trying to apply infinitesimals to geometry. Also, Desargues adopted an unfortunate and eccentric style of writing. He introduced some seventy new terms, many of a recondite botanical origin, of which only one, *involution*, has survived, and, curiously enough, this one was preserved because it was the piece of Desargues' technical jargon that was singled out for the sharpest criticism and ridicule by his reviewer.

167° *The precocity of Pascal.* Pascal early showed unusual ability in mathematics, and several stories of his youthful accomplishments have been told by his sister Gilberta, who became Madame Périer. Because of a delicate constitution, the boy was kept at home to insure his not being overworked. His father decided that the youngster's education should be at first restricted to the study of languages and should not include any mathematics. The exclusion of mathematics from his studies aroused curiosity in the boy and he inquired of his tutor as to the nature of geometry. The tutor informed him that it was the study of exact figures and the properties of their different parts. Stimulated by his tutor's description of the subject and by his father's injunction against it, he gave up his playtime and clandestinely, in a few weeks, discovered for himself many properties of geometrical figures, in particular the fact that the sum of the angles of a triangle is equal to a straight angle. This latter was accomplished by some process of folding a paper triangle, perhaps by folding the vertices over to the center of the inscribed circle, as indicated in Figure 24, or by folding the vertices over to the foot of an altitude, as indicated in Figure 25. When

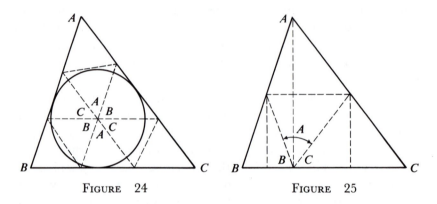

FIGURE   24   FIGURE   25

his father came upon him one day during his geometrical activities, he was so struck by the boy's ability that he gave his son a copy of Euclid's *Elements*, which the youngster read with avidity and quickly mastered.

At the age of fourteen, Pascal participated in the weekly gatherings of a group of French mathematicians from which the French Academy ultimately arose in 1666. When he was sixteen he wrote an essay on conic sections that Descartes could not believe was the work of the boy but thought it must be that of his father instead. At eighteen or nineteen he invented the first calculating machine, which he devised to assist his father in the auditing of government accounts at Rouen. Pascal was to manufacture over fifty calculating machines, some of which are still preserved in the Conservatoire des Arts et Métiers at Paris. At twenty-three he became interested in Torricelli's work on atmospheric pressure and began to apply his unusual talents to physics, with the result that *Pascal's principle* of hydrodynamics is today known to every student of high school physics.

168° *The greatest "might-have-been" in the history of mathematics.*
Pascal's astonishing and precocious activity came suddenly to an end
in 1650, when, suffering from frail health, he decided to abandon his
research in mathematics and science and to devote himself to religious
contemplation. Three years later, however, he returned briefly to
mathematics. At this time he wrote his *Traité du triangle arithmétique*
concerning a triangular arrangement of the binomial coefficients,
conducted several experiments on fluid pressure, and in correspondence
with Fermat assisted in laying the foundations of the mathematical
theory of probability. But late in 1654 he received what he regarded as
a strong intimation that these renewed activities were not pleasing to
God. The divine hint occurred when his runaway horses dashed over
the parapet of the bridge at Neuilly and he himself was saved only by
the miraculous breaking of the traces. Fortified with a reference to the
accident written on a small piece of parchment henceforth carried next
to his heart, he dutifully went back to his religious meditations.

Only once again, in 1658, did Pascal return to mathematics.
While suffering with toothache some geometrical ideas occurred to
him, and his teeth suddenly ceased to ache. Regarding this as a sign
of divine will, he obediently applied himself assiduously for eight days
toward developing his ideas, producing in this time a fairly full account
of the geometry of the cycloid curve and solving some problems that
subsequently, when issued as challenge problems, baffled other mathematicians. His famous *Provincial Letters* and his *Pensées*, which are read
today as models of early French literature, were written toward the
close of his brief life.

Pascal has been described as the greatest "might-have-been" in
the history of mathematics. With such unusual talents and such deep
geometrical intuition he should have produced, under more favorable
conditions, a great deal more. But his health was such that most of his
life was spent racked with physical pain, and from early manhood he
also suffered the mental torments of a religious neurotic.

169° *Pascal's "mystic hexagram" theorem.* The great gem of
Pascal's work on conic sections is the theorem, illustrated by Figure 26,
now known by his name: *The three points of intersection of the three pairs of
opposite sides of a (not necessarily convex) hexagon inscribed in a conic lie on a*

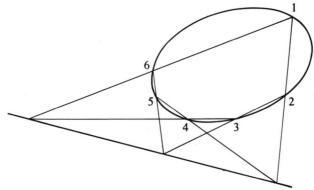

FIGURE   26

straight line. There is a report, quite likely a legend, that Pascal derived over four hundred corollaries to his great theorem, by considering special cases in which various of the vertices of the inscribed hexagon are allowed to coalesce. In any case, the consequences of Pascal's theorem are very numerous and attractive, and an almost unbelievable amount of research has been expended on the configuration. To illustrate, let us call the concerned line of collinearity the *Pascal line* of the given inscribed hexagon. There are $5!/2 = 60$ possible ways of forming a hexagon from six points on a conic, and, by Pascal's theorem, to each hexagon corresponds a Pascal line. These sixty Pascal lines pass three by three through twenty points, called *Steiner points*, which in turn lie four by four on fifteen lines, called *Plücker lines*. The Pascal lines also concur three by three in another set of points, called *Kirkman points*, of which there are sixty. Corresponding to each Steiner point, there are three Kirkman points such that all four lie upon a line, called a *Cayley line*. There are twenty of these Cayley lines, and they pass four by four through fifteen points, called *Salmon points*. There are many further extensions and properties of the configuration, and the number of different proofs that have been supplied for the "mystic hexagram" theorem itself is now legion.

170° *Lovis de Montalte and Amos Dettonville.* Pascal sometimes, as in his *Provincial Letters*, wrote under the nom de plume Lovis (Louis) de Montalte. In various of his challenge problems, he called himself Amos Dettonville, and this is why Leibniz occasionally refers to Pascal as Dettonville. The name Amos Dettonville is an anagram of the name Lovis de Montalte.

171° *Two very practical contributions.* It is interesting that Pascal has been credited with the invention of the one-wheeled wheelbarrow as we know it today. Also, at the age of thirty-five, he conceived the omnibus—an idea that was soon put into practice at five sous a ride.

172° *A specious use of probability.* In the seventh chapter of his *Pensées*, Pascal puts forth the argument that, since the value of eternal happiness must be infinite, then, even if the probability of a religious life ensuring happiness be very small, still the expectancy (which is measured by the product of the two) must be sufficient to render it worthwhile to be religious.

RENÉ DESCARTES AND PIERRE DE FERMAT

WHILE Desargues and Pascal were opening the new field of projective geometry, Descartes and Fermat were conceiving ideas of modern analytic geometry. There is a fundamental distinction between the two studies, for the former is a *branch* of geometry whereas the latter is a *method* of geometry. There are few academic experiences that can be more thrilling to the student of elementary mathematics than his introduction to this new and powerful method of attacking geometrical problems. The essence of the idea, as applied to the plane, is the establishment of a correspondence between ordered pairs of real numbers and points in the plane, thereby making possible a correspondence between curves in the plane and equations in two variables, so that for each curve in the plane there is an equation $f(x, y) = 0$, and for each such equation there is a curve, or set of points, in the plane. A correspondence is similarly established between the algebraic and analytic properties of the equation $f(x, y) = 0$ and the geometric properties of the associated curve. Geometry is cleverly reduced to algebra and analysis.

René Descartes was born near Tours in 1596. Shortly after he left school in 1612, he went to Paris and devoted some time to the study of mathematics. In 1617 he commenced several years of soldiering by first joining the army of Prince Maurice of Nassau. Upon quitting military life he spent four or five years traveling through Germany, Denmark, Holland, Switzerland, and Italy. After resettling for a couple of years in Paris, where he continued his mathematical studies and his philosophical contemplations, and where for a while he took up the construction of optical instruments, he decided to move to Holland, then at the height of its power. There he lived for twenty years, devoting his time to philosophy, mathematics, science, and writing. In 1649 he went to Stockholm, at the request of Queen Christina. He died there early in 1650.

Pierre de Fermat was born near Toulouse in 1601(?). He was the son of a leather merchant and received his early education at home. At the age of thirty he obtained the post of councillor for the local parliament at Toulouse and there discharged his duties with modesty and punctiliousness. Working as a humble and retiring lawyer, he devoted the bulk of his leisure time to the study of mathematics. Although he published very little during his lifetime, he was in scientific correspondence with many leading mathematicians of his day and in this way considerably influenced his contemporaries. He enriched so many branches of mathematics with so many important contributions that he has been called the greatest French mathematician of the seventeenth century. In particular, he is universally regarded as one of the very greatest number theorists of all times. He died in Castres in 1665.

173° *A challenge problem and a friendship.* There is a story, whose authenticity has been questioned, that after Descartes joined the army of Prince Maurice of Nassau in 1617, he was one day walking through the streets of Breda, where the army was stationed, when he came upon a posted placard in Dutch, which excited his curiosity. Stopping the first passerby, Descartes asked him to translate the placard into either French or Latin. The stranger, who happened to be Isaac Beeckman, the head of the Dutch College at Dort, agreed to do so provided Descartes would answer it, for the placard contained a challenge problem in geometry. Descartes solved the problem within a few hours and a warm friendship resulted between him and Beeckman.

174° *The birth of an idea.* There are a couple of legends describing the initial hint that led Descartes to the contemplation of analytic geometry. According to one story, it came to him in a dream. On St. Martin's Eve, November 10, 1619, while the army with which he was soldiering was lying inactive in its winter quarters on the banks of the Danube, Descartes experienced three singularly vivid and coherent dreams that, he claimed, changed the whole course of his life. The dreams, he said, clarified his purpose in life and determined his future endeavors by revealing to him "a marvelous science" and "a wonderful discovery." Descartes never explicitly disclosed just what were the marvelous science and the wonderful discovery, but some believe them to have been analytic geometry, or the application of algebra to geometry, and then the reduction of all the sciences to geometry. Eighteen years later he expounded some of his ideas in his famous philosophical treatise on universal science, *Discours de la méthode pour bien conduire sa raison et chercher la vérité dans les sciences* (Discourse on the Method of Correctly Guiding the Reason and Seeking Truth in the Sciences). This work was accompanied by three appendices illustrating the method; it was in the third of these that the world was given analytic geometry. The year of publication was 1637.

Another story, perhaps on a par with the story of Isaac Newton and the falling apple, says that the initial flash of analytic geometry came to Descartes when he was watching a fly crawl about on the ceiling near a corner of his room. It struck him that the path of the fly on the ceiling could be described if only one knew the relation connecting the fly's distances from the two adjacent walls. Even though this second story may be apocryphal, it has good pedagogical value.

175° *Descartes' advice.* As a youngster, Descartes suffered fragile health. When he was eight, he was sent to a Jesuit school at La Flèche. There Father Charlet, taking a personal interest in him and realizing that to educate the delicate boy's mind he must first build up his body, permitted the youngster to lie in bed in the mornings as late as he pleased and not to leave his room till he felt like joining his schoolmates. This pleasant custom developed into a lifelong habit. Descartes later averred that these meditative hours of morning rest constituted his most productive periods and were the source of his philosophy and mathematics.

When Descartes visited Pascal in 1647, he advised his sickly host to follow his example, to lie abed every morning till close to the noon hour and not to allow anyone to make him get up until he felt inclined to do so, claiming this to be the best way to do good work in mathematics and to preserve one's health. Unfortunately, Pascal did not heed his guest's excellent advice.

176° *Two significant contributions to mathematical notation.* In the third appendix, *La géométrie,* of his *Discours,* Descartes fixed the present-day custom of employing the first letters of the alphabet to denote known quantities and the last letters to denote unknown ones. He also introduced our present system of indices, such as a^3, a^4, and so forth. This system of indices had been suggested by earlier writers, but Descartes' adoption of the system, which was probably original on his part, fixed subsequent custom.

177° *The death of Descartes.* In 1649 Descartes received a pleading invitation from the nineteen-year-old Queen Christina of Sweden to bring learning to her court and to teach her philosophy. Dazzled by the aura of royalty and the man-of-war specially sent to fetch him, but yet with qualms and hesitation, Descartes finally exchanged his quiet and peaceful existence in Holland for a boisterous and hectic life in Stockholm. Queen Christina, who seems to have been much more of an athlete than a scholar, decided that five o'clock in the morning in a cold library with the windows thrown wide open was the only proper time and place to study philosophy. Accordingly, at some ungodly hour before the light of day, poor Descartes found himself roused from his warm bed and swept across a cold windy square to the chilly palace library to give the headstrong Queen her philosophy lessons. In addition to this madness, it was not very long before Descartes found himself the center of a veritable hornets' nest of malicious whisperings about foreign influence over the Queen. The cold unfriendly climate of Stockholm, the rupture of his deeply loved habit of lying abed until almost noon, and the destruction of his former quiet and private life were too much for him. After a few months in Sweden the weary philosopher fell ill with a fever accompanied by an inflammation of the lungs, dragged himself to a sick bed, and within ten days was dead, a victim of the conceited vanity of a willful girl.

178° *The Fermat numbers.* Some great men lead colorful personal lives that naturally give rise to fine stories and anecdotes, while others live such quiet, plodding, uneventful personal lives that nothing unusual is suggested. Pierre de Fermat belongs to the second category. And yet many stories can be told concerning Fermat, for though his personal life was indeed uneventful, his creative work in mathematics—or, in his case, we should say his recreation in mathematics—is full of interesting stories.

For example, Fermat died with the belief that he had found a long-sought-for prime-yielding expression in the formula

$$F_n = 2^{2^n} + 1.$$

In no less than seven different places did he express his conviction that F_n is a prime number for all nonnegative integral values of n, though he was always careful to admit that he had no proof of the conjecture. For $n = 0, 1, 2, 3, 4$ we have $F_n = 3, 5, 17, 257, 65{,}537$, respectively, and these are all prime numbers. But, in 1732, the great Swiss mathematician Leonhard Euler showed that

$$F_5 = 4{,}294{,}967{,}297 = (6{,}700{,}417)(641),$$

and thus is composite. Since then, by clever individual proofs not capable of any general application, F_n has also been shown to be composite for $n = 6, 7, 8, 9, 11, 12, 15, 18, 23, 36, 38, 73$, and in no other case than for $n = 0, 1, 2, 3, 4$ has F_n been shown to be prime. The result is that, whereas Fermat felt that F_n is prime for all nonnegative integral n, the general feeling among mathematicians today is that, except for $n = 0, 1, 2, 3,$ and 4, F_n is always composite.

The numbers F_n are known as *Fermat numbers*, and they increase rapidly in size with increasing n. For example F_{10}, the first case whose primality or nonprimality is not yet determined, contains 309 digits.

The number F_{36} has been shown to contain more than twenty trillion digits. And the number F_{73} has been shown to be so large that if it were written out in digits of type of the size in this book, all the books in all the libraries of the world would not suffice to record it!

A very remarkable discovery was made in 1796 in connection with the Fermat numbers. The ancient Greeks had shown how, with straight-edge and compasses, regular polygons of three, four, five, six, and fifteen sides can be constructed. By successive angle, or arc, bisections, we may then with Euclidean tools construct regular polygons having 2^n, $3(2^n)$, $5(2^n)$, and $15(2^n)$ sides. Not until almost the nineteenth century was it known that any other regular polygons can be constructed with these limited tools. In 1796, the eminent German mathematician Carl Friedrich Gauss, when only nineteen years old, developed the theory that shows that a regular polygon having a *prime* number of sides can be constructed with Euclidean tools if and only if that number is a prime Fermat number. Thus, unknown to the Greeks, regular polygons of 17, 257, and 65,537 sides can be constructed with straightedge and compasses.

Many Euclidean constructions of the regular polygon of seventeen sides have been given. In 1832, Richelot published an investigation of the regular polygon of 257 sides, and a Professor Hermes of Lingen gave up ten years of his life to the problem of constructing a regular polygon of 65,537 sides. It has been said that it was Gauss's youthful discovery that a regular polygon of seventeen sides can be constructed with straightedge and compasses that made him decide to devote his life to mathematics. His pride in this discovery is evidenced by his request that a regular polygon of seventeen sides be engraved on his tombstone. Although this request was never fulfilled, such a polygon is found on the base of a monument to Gauss erected at his birthplace in Brunswick.

179° *Fermat's method of infinite descent.* [The following is adapted, with permission, from the article by Howard Eves, of the same title, that appeared in the Historically Speaking section of *The Mathematics Teacher*, March, 1960, pp. 195–196.]

Diophantus, the famous Greek number theorist of antiquity, wrote at least three mathematical works, of which his *Arithmetica* is by far the most important. The *Arithmetica* is an analytical treatment of algebraic number theory that marks the author as a genius in this field. Only six of the original thirteen books of the *Arithmetica* are extant, and these six books are devoted to the solution of about 130 number problems of considerable variety and degree of difficulty. There have been many commentaries of the *Arithmetica*, but it was Regiomontanus who, in 1463, called for a Latin translation of the extant Greek text. The call was met in 1575 by Xylander (the Greek name assumed by Wilhelm Holzmann, a professor of mathematics at the University of Heidelberg), who made a meritorious translation with commentary. This Xylander translation was later used by the Frenchman Bachet de Méziriac when, in 1621, he published the first edition of the Greek text along with a Latin translation and notes.

Pierre de Fermat secured a copy of the Bachet edition of Diophantus's *Arithmetica* and used the copy as a combination textbook and notebook. Many of Fermat's contributions to the field of number theory occur as marginal statements inserted in his copy of the *Arithmetica*. In 1670, five years after Fermat's death, these marginal notes were incorporated in a new, but unfortunately carelessly printed, edition of the *Arithmetica*, brought out by Fermat's son, Clément-Samuel. In this edition we find, accompanying Problem 26 of Book VI, the following marginal note by Fermat.

> The area of a right triangle whose sides are rational numbers cannot be a square number. I have obtained a proof of this theorem only after extensive and arduous effort. I here reproduce the proof, since the procedure used will make possible wonderful progress in number theory.

Then follows an indication of the proof.

The procedure alluded to by Fermat is ingenious, and has since become known as *Fermat's method of infinite descent*. The method apparently was used with success by Fermat on a number of occasions. For example, in one of his letters to Roberval, Fermat describes the difficulties he experienced in trying to establish a celebrated conjecture made by Bachet that every positive integer can be written as the sum of at most four squares, and in the letter Fermat says that he finally succeeded by the use of his favorite method of infinite descent. Again, in 1897 a paper was found in the library at Leyden among the manuscripts of Christiaan Huygens, in which Fermat describes the method of infinite descent.

The method of infinite descent is particularly useful in establishing negative results. In outline the method is this. To prove that there do not exist positive integers $a, b, c, \ldots$ satisfying a relation $R(a, b, c, \ldots)$, assume the contrary. On this assumption show that $R(a_1, b_1, c_1, \ldots)$ holds, where a_1 is a positive integer and $a_1 < a$. Then in like manner we may show that $R(a_2, b_2, c_2, \ldots)$ holds, where a_2 is a positive integer and $a_2 < a_1$, and so on ad infinitum. But, since there are only a finite number of positive integers less than a, this is impossible. We are thus led to a contradiction, whence we conclude that the relation $R(a, b, c, \ldots)$ is not satisfied by positive integers $a, b, c, \ldots$.

To clarify the Fermat method of infinite descent, let us consider a simple application of the method. Let us prove by the method that $\sqrt{2}$ is irrational. Suppose, on the contrary, that $\sqrt{2} = a/b$, where a and b are positive integers. Now

$$\sqrt{2} + 1 = 1/(\sqrt{2} - 1),$$

whence

$$\frac{a}{b} + 1 = \frac{1}{\dfrac{a}{b} - 1} = \frac{b}{a - b},$$

and

$$\sqrt{2} = \frac{a}{b} = \frac{b}{a - b} - 1 = \frac{2b - a}{a - b} = \frac{a_1}{b_1}, \text{ say.}$$

But, since $1 < \sqrt{2} < 2$, after replacing $\sqrt{2}$ by a/b and then multiplying through by b, we have $b < a < 2b$. Now, since $a < 2b$, it follows that $0 < 2b - a = a_1$. And since $b < a$, it follows that $a_1 = 2b - a < a$. Thus a_1 is a positive integer less than a. By a reapplication of our procedure we find $\sqrt{2} = a_2/b_2$, where a_2 is a positive integer less than a_1. The process may be repeated indefinitely. But the positive integers cannot be decreased in magnitude indefinitely. It therefore follows that our original assumption that $\sqrt{2} = a/b$, where a and b are positive integers, is untenable. That is, $\sqrt{2}$ is irrational.

180° *The most tantalizing marginal note in the history of mathematics.*   Of the well over three thousand mathematical papers and notes that he wrote, Fermat published only one, and that just five years before his death and under the concealing initials M. P. E. A. S. Many of his mathematical findings were disclosed in letters to fellow mathematicians and in marginal notes inserted in his copy of Bachet's translation of Diophantus's *Arithmetica*.

At the side of Problem 8 of Book II in his copy of Diophantus, Fermat wrote what has become the most tantalizing marginal note in the history of mathematics. The considered problem in Diophantus is: "To divide a given square number into two squares." Fermat's accompanying marginal note reads:

> To divide a cube into two cubes, a fourth power, or in general any power whatever above the second, into two powers of the same denomination, is impossible, and I have assuredly found an admirable proof of this, but the margin is too narrow to contain it.

This famous conjecture, which says that *there do not exist positive integers x, y, z, n such that* $x^n + y^n = z^n$ *when n > 2*, has become known as "Fermat's last theorem." Whether Fermat really possessed a sound demonstration of this conjecture will probably forever remain an enigma. Because of his unquestionable integrity we must accept as a fact that he thought he had a proof, and because of his paramount ability we must accept as a fact that if the proof contained a fallacy then that fallacy must have been very subtle.

Many of the most prominent mathematicians since Fermat's time have tried their skill on the problem, but the general conjecture still remains open. There is a proof given elsewhere by Fermat for the case $n = 4$, and Euler supplied a proof (later perfected by others) for $n = 3$. About 1825, independent proofs for the case $n = 5$ were given by Legendre and Dirichlet, and in 1839 Lamé proved the conjecture for $n = 7$. Very significant advances in the study of the problem were made by the German mathematician E. Kummer. In 1843, Kummer submitted a purported proof of the general conjecture to Dirichlet, who pointed out an error in the reasoning. Kummer then returned to the problem with renewed vigor, and a few years later, after developing an important allied subject in higher algebra called the *theory of ideals*, derived very general conditions for the insolvability of the Fermat relation. Almost all important subsequent progress on the problem has been based on Kummer's investigations. It is now known that "Fermat's last theorem" is certainly true for all $n < 4003$ (this was shown in 1955, with the aid of the SWAC digital computer), and for many other special values of n.

In 1908, the German mathematician P. Wolfskehl bequeathed 100,000 marks to the Academy of Science at Göttingen as a prize for the first complete proof of the "theorem." The result was a deluge of alleged proofs by glory-seeking and money-seeking laymen, and, ever since, the problem has haunted amateurs somewhat as does the trisection of an arbitrary angle and the squaring of the circle. "Fermat's last theorem" has the peculiar distinction of being the mathematical problem for which the greatest number of incorrect proofs have been published.

In Mathematical Circles

Quadrants III and IV

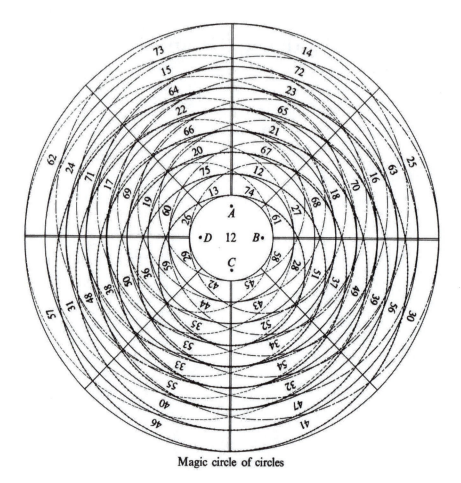

Magic circle of circles

The "magic circle of circles," shown in this volume, was created by
Benjamin Franklin, and the intricacies of this construction are unfolded
in Item 317°.

In Mathematical Circles

Quadrants III and IV

Howard Eves

Published and Distributed by
The Mathematical Association of America

CONTENTS

QUADRANT THREE

CONTENTS

Contents

QUADRANT FOUR

CONTENTS

CONTENTS

QUADRANT THREE

*From some arguments of priority
to a doubtfully credited theorem*

SOME MINOR STORIES ABOUT SOME MINOR MEN

BEFORE considering Isaac Newton and the period following him, we here dispose of a few stories about some of Newton's lesser contemporaries.

181° *Gilles Persone de Roberval and arguments of priority.* Gilles Persone was born at Roberval, France, in 1602 and died in Paris in 1675. He assumed the seignioral title of de Roberval, to which he was not entitled. His extensive correspondence served as a medium for the intercommunication of mathematical ideas in those pre-journal days. He became well known for his method of drawing tangents and his discoveries in the field of higher plane curves. The idea of tangents was also held by Torricelli, and an argument of priority ensued. Roberval also claimed to be the inventor of Cavalieri's pre-calculus method of indivisibles and to have squared the cycloid before Torricelli.

The above matters of priority are difficult to settle, for Roberval was consistently tardy in disclosing his discoveries. This tardiness has been explained by the fact that for forty years, starting in 1634, Roberval held a professorial chair at the Collège Royale. This chair automatically became vacant every three years, to be filled by open competition in a mathematical contest in which the questions were set by the outgoing incumbent. To retain his position, Roberval saved his discoveries to use as difficult contest questions that he would be able to answer but that his competitors would probably find troublesome.

182° *John Wallis and the invention of the infinity symbol.* John Wallis was born at Ashford, England, in 1616 and died at Oxford in 1703. Wallis was one of the ablest and most original mathematicians of his day. He was a voluminous and erudite writer in a number of fields and is said to have been one of the first to devise a system for teaching deaf-mutes. In 1649 he was appointed Savilian professor of geometry at Oxford, a position that he held for fifty-four years until his death. His work in analysis did much to prepare the way for his great contemporary, Isaac Newton. He was one of the founders of the Royal Society and for years assisted the government as a cryptologist.

3

Wallis was the first to use the symbol ∞ for infinity, in his *Arithmetica infinitorum* of 1655, but it was not generally adopted until Jakob Bernoulli used it in 1713 in his *Ars conjectandi*. The symbol had sometimes been used by the Romans to denote the number 1000, and it may be that this led to its being employed to represent any very large number, from which it then graduated to represent infinity.

183° *A facetious remark about the symbol for infinity.* There is a saying that numbers don't lie, but this one does—it is a number 8 lying on its side.

184° *Nicolaus Mercator and the loss of a fee.* Nicolaus Mercator was born at Holstein, then a part of Denmark, about 1620 and died at Paris in 1687. He edited Euclid's *Elements* and wrote on trigonometry, astronomy, the computation of logarithms, and cosmography. He lived most of his life in England but went to France in 1683, where he designed and constructed the fountains at Versailles. The payment agreed upon for this work was refused him unless he would turn Catholic. This he refused to do and shortly after died of frustration and poverty.

The familiar map of a sphere known as *Mercator's projection*, in which loxodromes (the paths on a sphere that make a constant angle with the meridians) appear as straight lines, is not due to Nicolaus Mercator, but to Gerard Mercator (1512–1594).

185° *Isaac Barrow: prankster, wit, strong man, generous teacher.* Isaac Barrow was born in London in 1630 and died in Cambridge in 1677. He accomplished much during his relatively short life and in some ways may be thought of as the Eratosthenes of his time. Like Eratosthenes, he was an outstanding athlete as well as a many-sided scholar—excelling in such diversified fields as mathematics, Greek, physics, and theology. Again like Eratosthenes, though he was gifted in many areas of knowledge, he failed (except perhaps in Greek) to top his contemporaries in any one branch. For some five years he traveled extensively in eastern Europe. In 1664 he was elected the first Lucasian professor at Cambridge. Entertaining stories are told of

his playful mischievousness, ready wit, great physical strength, coura-
geous bravery, and scrupulous conscientiousness.

Barrow's playful mischievousness and its occasional irritability are
brought out in a story dating from his early school days. One evening
at that time he overheard his exasperated father praying, "Lord, if
ever for some reason you should decide to take one of my children, I
could best spare Isaac."

Barrow's ready wit made him a favorite of Charles II. This wit is
illustrated by an incident that has been reported by W. W. R. Ball.
During Barrow's examination for holy orders, the dialogue is said to
have run as follows:

Chaplain. *Quid est fides?*
Barrow. *Quod non vides.*
Chaplain. *Quid est spes?*
Barrow. *Magna res.*
Chaplain. *Quid est caritas?*
Barrow. *Magna raritas.*

The chaplain, shocked at his candidate's rhyming irreverence, here
gave up in despair and reported Barrow to the bishop. But the bishop,
fortunately, had a sense of humor and Barrow was duly installed.

Though Barrow was short, lean, pale, and an inveterate smoker,
it seems that he could have posed as the strong man in any circus. It
is said that he could tie iron rods into knots and then untie them, and
that he was a surprisingly powerful swimmer. Once, in his travels in
eastern Europe, his prowess and courage saved the ship he and some
other passengers were sailing on from capture by pirates.

In 1669, in a fine and laudable gesture, Barrow magnanimously
stepped down from the Lucasian chair at Cambridge in favor of his
great pupil, Isaac Newton, whose remarkable abilities he was one of
the first to recognize and acknowledge.

186° *Sir Christopher Wren and the Great London Fire.* Occasionally an incident occurs in a man's life that causes his fame to posterity to shift from an anticipated field to an unanticipated one. This
happened to Sir Christopher Wren (1632–1723), for it seems that but
for London's Great Fire of 1666, Wren would have been known as a
mathematician instead of as an architect. He was Savilian professor of

5

astronomy at Oxford from 1661 to 1673, and, for a time, president of the Royal Society. He wrote on the laws of collision of bodies, on subjects connected with optics, the resistance of fluids, and other topics in mathematical physics and celestial mechanics. He is credited with the discovery, in 1669, of the two systems of rulings on the hyperboloid of one sheet. He was the first (1658) to show that an arch of the cycloid is equal in length to eight times the radius of the generating circle. But after the Great Fire, Wren took such a prominent part in the rebuilding of St. Paul's Cathedral and some fifty or more other churches and public buildings that his fame as an architect overshadowed his reputation as a mathematician.

PRE-NEWTONIAN VERSUS POST-NEWTONIAN MATHEMATICS

Mathematicians have found various significant differences between the older and the newer mathematics. Some have found the major difference to lie in the growing *interconnectedness* of the subject. Others have found the major difference to lie in the changing *methods* of the subject. And still others have found the major difference to lie in the altering *viewpoints* of the subject.

187° *Mathematics as a rising continent.* The development of mathematics over the ages may be viewed as a continent slowly rising from the sea. At first perhaps a single island appears, and, as it grows in size, other islands emerge at varying distances from one another. As the continent continues to rise, some of the islands become joined to others by isthmuses that widen until pairs of islands become single large islands. At length a point is reached where the shape of the continent is essentially defined, and there remain only a number of lakes and inland seas of various sizes. As the continent further rises, these lakes and seas shrink and vanish one by one. The older mathematics compares to the situation when the general shape of the rising continent is still undefined and the land area consists largely of islands of different sizes. The newer mathematics compares to the situation when the general shape of the rising continent has become essentially clear, with most of the former islands now joined by stretches of land.

188° *Mathematics as a rock.* Mathematics may be likened to a large rock whose interior composition we wish to examine. The older mathematicians appear as persevering stone cutters slowly attempting to demolish the rock from the outside with hammer and chisel. The later mathematicians resemble expert miners who seek vulnerable veins, drill into these strategic places, and then blast the rock apart with well placed internal charges.

189° *The finite and the infinite in mathematics.* Older mathematics appears static while the newer appears dynamic, so that the older mathematics compares to the still-picture stage of photography while the newer mathematics compares to the moving-picture stage. Again, the older mathematics is to the newer much as anatomy is to physiology, wherein the former studies the dead body and the latter studies the living body. Once more, the older mathematics concerned itself with the fixed and the finite while the newer mathematics embraces the changing and the infinite.

ISAAC NEWTON AND GOTTFRIED WILHELM LEIBNIZ

ISAAC Newton was born in Woolsthorpe on Christmas Day, 1642, the year in which Galileo died. His father, who died before Isaac was born, was a farmer, and it was planned that the son also should devote his life to farming. The youngster, however, showed great skill and delight in devising clever mechanical models and in conducting experiments. The result was that his schooling was extended, and, when eighteen, he was allowed to enter Trinity College at Cambridge. During 1665, and part of the following year, the university closed because of the rampant bubonic plague, and Newton stayed home. It was during this period that he developed his calculus, became interested in various physical questions, and formulated the basic principles of his theory of gravitation. He returned to Cambridge in 1667 and for two years occupied himself chiefly with optical researches. In 1669, his teacher Isaac Barrow resigned the Lucasian professorship in his favor, and Newton began his eighteen years of university teaching. It was during this time that, among other things, Newton developed the material of his great *Principia,* which was published, at his friend Halley's expense, in the middle of 1687 and immediately made an enormous impression throughout Europe. In 1692, Newton suffered a curious illness that lasted about two years and that involved some form of mental derangement. Most of his later life was devoted to chemistry, alchemy, and theology. In 1696 he was appointed Warden of the Mint, and in 1699 he was promoted to Master of the Mint. In 1703 he was elected president of the Royal Society, a position to which he was annually re-elected until his death, and in 1705 he was knighted. The last part of his life was made unhappy by the unfortunate controversy with Leibniz concerning incriminations of plagiarism in the discovery of the calculus. He died in 1727 at the age of eighty-four, after a lingering and painful illness, and was buried in Westminster Abbey. At death he had all of his teeth except one, and a full head of white hair. History has recorded him as perhaps the greatest mathematician of all time.

Gottfried Wilhelm Leibniz, the great universal genius of the seventeenth century and Newton's rival in the invention of the calculus, was born in Leipzig in 1646. Before he was twenty he began to develop the first ideas of his "characteristica generalis," which involved a universal mathematics that later blossomed into the symbolic logic of George Boole (1815–1864), and still later, in 1910, into the *Principia mathematica* of Whitehead and Russell. When, ostensibly because of his youth, he was refused the degree of doctor of laws at the University of Leipzig, he moved to Nuremberg. There he wrote a brilliant essay on teaching law by the historical method and dedicated it to the Elector of Mainz. This led to his appointment by the Elector to a commission for the recodification of some statutes. The rest of Leibniz's life from this point on was spent in diplomatic service, first for the Elector of Mainz, and then from about 1676 until his death, for the estate of the Duke of Brunswick at Hanover. In 1672, while in Paris on a diplomatic mission, he met the great Dutch mathematician Huygens, who was then residing there, and prevailed upon the scientist to give him lessons in mathematics. The following year he was sent on a political mission to London, where he exhibited a calculating machine he had invented to the Royal Society. Before he left Paris to take up his lucrative post as librarian for the Duke of Brunswick, Leibniz had already discovered the fundamental theorem of the calculus, developed much of his notation in this subject, and worked out a number of the elementary formulas of differentiation. In addition to being a skilled mathematician, Leibniz was a particularly gifted linguist, winning some fame as a Sanskrit scholar, and his writings on philosophy have ranked him high in that field. In 1682, he and Otto Mencke founded a journal called the *Acta eruditorum*, of which he became editor-in-chief. In 1700 he founded the Berlin Academy of Science, and endeavored to create similar academies in Dresden, Vienna, and St. Petersburg. The closing years of his life were embittered by the controversy that others brought upon him and Newton concerning whether he had discovered the calculus independently of Newton. In 1714, his employer became the first German King of England, and Leibniz was left, neglected, at Hanover. Two years later, in 1716, when he died, his funeral was attended only by his faithful secretary.

190° *An ingenious young inventor.*   Isaac Newton amused himself in his boyhood by devising a wide array of gadgets and mechanical toys. Thus he constructed waterwheels, windmills, sundials, a wooden clock that worked by water, and many perfectly constructed toys for friends. He made a toy gristmill that ground wheat to flour, with a mouse serving as motive power, and he constructed lantern-carrying kites to affright the gullible villagers at night, who thought the clever things were comets.

191° *Genius is not always bright.*   There is a story about Newton as a youngster that points up how even a genius may at times be a bit stupid. Isaac was sent to the barn to cut a hole at the bottom of the barn door for the cats to go in and out of. He cut two holes, a large one for the cat and a small one for the kittens.

192° *Spurred on by a bully.*   All through his life, Newton seems to have required an external stimulus to rouse his latent power and to make him exert himself to complete his work or to render his findings public. An early story of this propensity is told of Newton when he was a schoolboy at Grantham, where at the start he was very negligent in his studies. When he was last in the lowermost class but one, the boy next above him, one day as they were going to school, kicked Newton in the stomach, causing him much physical pain and mental anguish. When school was over, Newton challenged the bully to a fight in the churchyard and roundly defeated him. Determined to beat his adversary also at the books, Newton, by hard work, finally succeeded, and then gradually rose until he was the top scholar in the whole school.

193° *Introduction to mathematics.* In 1661, Newton entered Trinity College at Cambridge. It was not until this stage of his schooling that his attention came to be directed to mathematics. At the beginning of his first October term, Newton happened to stroll down to Stourbridge Fair, where he picked up a book on astrology. Unable to understand the book because of the geometry and trigonometry involved, he bought a copy of Euclid's *Elements*, which he found to be very easy, and then Descartes' *Géométrie*, which he found somewhat difficult. The interest in mathematics that these books engendered in him led him to take up mathematics rather than chemistry as a serious study, and he went on and read Oughtred's *Clavis*, works of Kepler and Viète, and the *Arithmetica infinitorum* of Wallis. From reading mathematics, he then turned to creating it.

194° *Newton's absent-mindedness.* It has been reported that Newton often spent eighteen or nineteen hours of the day in writing, and that he possessed remarkable powers of concentration. Amusing tales, perhaps apocryphal, are told in support of his absent-mindedness when engaged in thought.

Thus there is the story that related that, when giving a dinner to some friends, Newton left the table for a bottle of wine, and becoming mentally engaged he forgot his errand, went to his room, donned his surplice, and ended up in chapel.

On another occasion, Newton's friend Dr. Stukeley called on him for a chicken dinner. Newton was out for the moment, but the table was already laid with the cooked fowl in a dish under a cover. Dr. Stukeley lifted the cover, removed and ate the chicken, and then replaced the bones in the covered dish. When Newton later appeared he greeted his friend and sitting down he, too, lifted the cover, only to discover the remains. "Dear me," he said, "I had forgotten that we had already dined."

And then there was the occasion when, riding home one day from Grantham, Newton dismounted from his horse to walk the animal up Spittlegate Hill just beyond the town. On the way up the hill the horse slipped away leaving only the empty bridle in his master's hands, a fact that Newton discovered only when, at the top of the hill, he endeavored to vault into the saddle.

195° *Newton's first purely scientific experiment.* Newton used to say, in his old age, that he performed his first purely scientific experiment on the day of Cromwell's death when he was sixteen years old. On that day a great storm raged over all England. To ascertain the strength of the wind he first broad-jumped with the wind and then against the wind. Comparing these distances with the extent of his broad-jump on a calm day, he secured a measure of the strength of the wind. Thereafter, when the wind blew, he used to measure it as so many feet strong.

196° *Some tributes paid to Newton.* Newton was a skilled experimentalist and a superb analyst. As a mathematician, he is ranked almost universally as the greatest the world has yet produced. His insight into physical problems and his ability to treat them mathematically has probably never been excelled. One can find many testimonials by competent judges to his greatness, such as the noble tribute paid by Leibniz, who said, "Taking mathematics from the beginning of the world to the time when Newton lived, what he did was much the better half." And there is the remark by Lagrange to the effect that Newton was the greatest genius that ever lived, and the most fortunate, for we can find only once a system of the universe to be established. His accomplishments were poetically expressed by Alexander Pope in the lines,

> Nature and Nature's laws lay hid in night;
> God said, "Let Newton be," and all was light.

Havelock Ellis described Newton as "the supreme respresentative of Anglo-Saxon genius." For other great mathematicians, Gauss used the epithets "magnus," or "clarus," or "clarissimus"; for Newton alone he kept the prefix "summus."

197° *Newton's estimate of himself.* In contrast to the eulogies of others is Newton's own modest estimate of his work: "I do not know what I may appear to the world; but to myself I seem to have been only like a boy playing on the seashore, and diverting myself in now and then finding a smoother pebble or a prettier shell than ordinary, whilst the great ocean of truth lay all undiscovered before me." In generosity to his predecessors he once explained: "If I have seen a little farther than others it is because I have stood on the shoulders of giants." He would not admit that there was any difference between himself and others, except perhaps in the passion of perseverance and vigilance. When asked how he made his discoveries, he replied: "By always thinking about them." He once declared that if he had accomplished anything, it was due solely to industry and patient thought: "I keep the subject of my inquiry constantly before me, and wait till the first dawning opens gradually, by little and little, into a full and clear light."

198° *De Morgan's anagrams on Newton.* Actually, others had predicted the inverse square law of gravitation* before Newton, but Newton added to it, verified it, and derived consequences from it. Augustus De Morgan summed up the situation by saying: "The notion of gravitation was *not new*; but Newton *went on*."

199° *Newton's dislike of controversy.* A number of Newton's discoveries, such as his theory of light and certain deductions from his optical experiments, were vehemently attacked by some scientists. Newton found the ensuing arguments, which soon degenerated from the purely scientific to the carping and the outright stupid, so distasteful that he vowed never to publish anything on science again. He was bewildered that scientific men should convert science into a battle-

---

* Any two particles in the universe attract one another with a force which is directly proportional to the product of their masses and inversely proportional to the square of the distance between them.

ground for personal quarrels, and he built up such a tremendous dislike of controversy that it seemed to border on the pathological. This had an important bearing on the history of mathematics, for the result was that almost all his findings remained unpublished until many years after their discovery. This postponement of publication later led to the undignified dispute with Leibniz concerning incriminations of plagiarism in connection with the discovery of the calculus. It was owing to this latter controversy that the English mathematicians, backing Isaac Newton as their leader, cut themselves off from continental developments, and mathematical progress in England was detrimentally retarded for almost a hundred years.

Writing on December 9, 1675, in connection with the voluminous correspondence he was dragged into in order to defend his theory of light, Newton said: "I was so persecuted with discussions arising out of my theory of light, that I blamed my own imprudence for parting with so substantial a blessing as my quiet to run after a shadow." And later, on November 18, 1676, he wrote: "I see I have made myself a slave to philosophy [science]; but if I get rid of Mr. Linus's business [Mr. Linus was one of the carping critics], I will resolutely bid adieu to it eternally, excepting what I do for my private satisfaction, or leave to come out after me; for I see a man must either resolve to put out nothing new, or to become a slave to defend it." Newton's exasperation at futile controversies broke out again after the publication of his *Principia*. Writing to Edmund Halley on June 20, 1688, he said: "[Science] is such an impertinently litigious Lady, that a man had as good be engaged in lawsuits, as to have to do with her. I found it so formerly, and now I am no sooner come near her again, but she gives me warning."

200° *The challenge problems.* In the squabble between England and the Continent as to the relative proficiency of the Newtonian method of fluxions and the Leibnizian method of differentials, and according to the custom of the day, challenge problems were proposed to test the two approaches. Thus, in June of 1696, Johann Bernoulli and Leibniz jointly proposed two challenge problems "to the acutest mathematicians of the world." The better of the two problems was: to determine the curve connecting two given points, at different heights and not in the same vertical line, along which a particle will slide (without friction) under the influence of gravity from the upper point to the lower point in the least possible time. This is the famous *brachisto-chrone* ("shortest time") problem, and the solution turns out to be an inverted cycloidal arc. The problem, and its companion problem, baffled the mathematicians of Europe for the whole six-month period of the contest. Accordingly, on the recommendation of Leibniz, the time was publicly extended for a year longer. Someone thereupon sent the two problems to Newton, who in this way heard of them for the first time on January 26, 1697, at four o'clock in the afternoon after a tiring day at the Mint. Sensing that the problems were sent to him to test the claims on behalf of the method of fluxions, he sat down after dinner and solved the two problems, and on the following day he anonymously communicated his solutions to the Royal Society. Upon seeing the solutions, Bernoulli is said to have at once exclaimed, "Ah, I recognize the lion by his paw." Newton's incredible feat proved that his unmatchable mathematical powers were, at the age of fifty-five, as vigorous as ever.

A further proof of Newton's continued mathematical vitality came in 1716, when he was seventy-four. Leibniz proposed what he considered to be a difficult problem, as a challenge to the mathematicians of Europe, but, without doubt, aimed particularly at Newton. The problem was to find the orthogonal trajectories of a one-parameter family of curves. As before, Newton received the problem late in the afternoon upon returning home exhausted from his day at the Mint. And again he solved the problem the very evening he received it. Newton appears to have had no equal, throughout the whole history of mathematics, in his ability to concentrate all the powers of his intellect on a great difficulty at a moment's notice.

201° *Leibniz's unusual mind.*   There are two broad and antithetical domains of mathematical thought, the continuous and the discrete. Leibniz is the one man in the history of mathematics who possessed both of these qualities of thought to a superlative degree.

202° *Leibniz and religion.*   Leibniz entertained various grand projects for religion that came to nought. Thus he attempted to reunite the Protestant and Catholic churches, and, when that effort failed, just the two Protestant sects of his day. E. T. Bell has pointed out that Leibniz overlooked a law that is as basic in religion as the second law of thermodynamics is in physics, namely that each religious sect tends to split into two, and then each of these into two more, and so on, until there are fewer human beings than there are creeds.

On another occasion, Leibniz felt he might have a way of Christianizing all of China by what he believed to be the image of creation in the binary arithmetic. Since God may be represented by unity, and nothing by zero, he imagined that God created everything from nothing just as in the binary arithmetic all numbers are expressed by means of unity with zero. This idea so pleased Leibniz that he communicated it to the Jesuit Grimaldi, President of the Mathematical Board of China, with the hope that it might convert the reigning Chinese emperor (who was particularly attached to science), and thence all of China, to Christianity.

THE BERNOULLIS

THERE is a general rule to the effect that any given family possesses at most one outstanding mathematician and that, in fact, most families possess none. Thus a search through the ancestors, descendants, and relatives of Isaac Newton fails to turn up any other great mathematician. There are exceptions to this general rule. For example we have, here in the United States, the two Lehmers (father and son), the two Birkhoffs (father and son), and some teams of married mathematicians. One also recalls the Italian-French Cassini family of the seventeenth and eighteenth centuries, and perhaps one can build a case for the Gregorys (uncle and nephew) of seventeenth-century England, the Van Schooten family of seventeenth-century Netherlands, and the Clairaut family of eighteenth-century France. And of course there were Theon and Hypatia (father and daughter), who lived during the closing years of ancient Greek mathematics. One can think of a few others. But such cases are relatively rare. All the more striking, then, is the Bernoulli family of Switzerland, which in three successive generations produced no less than eight noted mathematicians.

203° *The Bernoulli family.* The Bernoulli family record starts
with the two brothers Jakob Bernoulli (1654–1705) and Johann Bernoulli (1667–1748). These two men gave up earlier vocational interests
and became mathematicians when Leibniz's papers began to appear
in the *Acta eruditorum.* They were among the first mathematicians to
realize the surprising power of the calculus and to apply the tool to a
great diversity of problems. From 1687 until his death, Jakob occupied
the mathematical chair at Basel University. Johann, in 1697, became
a professor at Groningen University, and then, on Jakob's death in
1705, succeeded his brother in the chair at Basel University, to remain
there for the rest of his active life. The two brothers, often bitter
rivals, maintained an almost constant exchange of ideas with Leibniz
and with each other.

Among Jakob Bernoulli's contributions to mathematics are the
early use of polar coordinates, the derivation in both rectangular and
polar coordinates of the formula for the radius of curvature of a plane
curve, the study of the catenary curve with extensions to strings of
variable density and strings under the action of a central force, the
study of a number of other higher plane curves, the discovery of the
so-called *isochrone*—or curve along which a body will fall with uniform
vertical velocity (it turned out to be a semicubical parabola with a
vertical cusptangent), the determination of the form taken by an
elastic rod fixed at one end and carrying a weight at the other, the
form assumed by a flexible rectangular sheet having two opposite
edges held horizontally fixed at the same height and loaded with a
heavy liquid, and the shape of a rectangular sail filled with wind. He
also proposed and discussed the problem of isoperimetric figures
(planar closed paths of given species with fixed perimeter which include
a maximum area), and was thus one of the first mathematicians to work
in the calculus of variations. He was also one of the early students of
mathematical probability; his book in this field, the *Ars conjectandi*,
was posthumously published in 1713.

There are several things in mathematics that now bear Jakob
Bernoulli's name. Among these are the *Bernoulli distribution* and *Bernoulli
theorem* of statistics and probability theory, the *Bernoulli equation* met
by every student of a first course in differential equations, the *Bernoulli
numbers* and *Bernoulli polynomials* of number-theory interest, and the

lemniscate of Bernoulli encountered in any first course in the calculus. In Jakob Bernoulli's solution to the problem of the isochrone curve, which was published in the *Acta eruditorum* in 1690, we meet for the first time the word "integral" in a calculus sense. Leibniz had called the integral calculus *calculus summatorius*; in 1696 Leibniz and Johann Bernoulli agreed to call it *calculus integralis*.

Johann Bernoulli was an even more prolific contributor to mathematics than his brother Jakob. Though he was a jealous and cantankerous man, he was one of the most successful teachers of his time. He greatly enriched the calculus and was very influential in making the power of the new subject appreciated in continental Europe.

Johann Bernoulli wrote on a wide variety of topics, including optical phenomena connected with reflection and refraction, the determination of the orthogonal trajectories of families of curves, rectification of curves and quadrature of areas by series, analytical trigonometry, the exponential calculus, and other subjects. One of his more noted pieces of work is his contribution to the problem of the *brachistochrone*— the determination of the curve of quickest descent of a weighted particle moving between two given points in a gravitational field; the curve turned out to be an arc of an appropriate cycloid curve. This problem was also discussed by Jakob Bernoulli. The cycloid curve is also the solution to the problem of the *tautochrone*—the determination of the curve along which a weighted particle will arrive at a given point of the curve in the same interval no matter from what initial point of the curve it starts. This latter problem, which was more generally discussed by Johann Bernoulli, Euler, and Lagrange, had earlier been solved by Huygens (1673) and Newton (1687), and applied by Huygens in the construction of pendulum clocks.

Johann Bernoulli had three sons, Nicolaus (1695–1736), Daniel (1700–1782), and Johann (II) (1710–1790), all of whom won renown as eighteenth-century mathematicians and scientists. Nicolaus, who showed great promise in the field of mathematics, was called to the St. Petersburg Academy, where he unfortunately died—by drowning —only eight months later. He wrote on curves, differential equations, and probability. A problem in probability, which he proposed from St. Petersburg, later became known as the *Petersburg paradox*. The problem is: if *A* will receive a penny should heads appear on the first toss of a coin, 2 pennies if heads does not appear until the second toss, 4 pennies if heads does not appear until the third toss, and so on, what is A's expectation? Mathematical theory shows that A's expectation is infinite, which seems a paradoxical result.

The Petersburg paradox problem was investigated by Nicolaus's brother Daniel, who succeeded Nicolaus at St. Petersburg. Daniel returned to Basel seven years later. He was the most famous of Johann's three sons, and devoted most of his energies to probability, astronomy, physics, and hydrodynamics. In probability he devised the concept of *moral expectation*, and in his *Hydrodynamica*, of 1738, appears the principle of hydrodynamics that bears his name in all present-day elementary physics texts. He wrote on tides, established the kinetic theory of gases, studied the vibrating string, and pioneered in partial differential equations. He was awarded the prize of the French Academy no fewer than ten times.

Johann (II), the youngest of Johann Bernoulli's three sons, studied law but spent his later years as a professor of mathematics at the University of Basel, succeeding his father in that position in 1743. He was particularly interested in the mathematical theory of heat and light. He received the prize of the French Academy on three occasions.

There was another eighteenth-century Nicolaus Bernoulli (1687–1759), a nephew of Jakob and Johann, who achieved some fame in mathematics. This Nicolaus held, for a time, the chair of mathematics at Padua once filled by Galileo. He wrote extensively on geometry, differential equations, infinite series, and probability. Later in life he taught logic and law.

Johann Bernoulli (II) had three sons, Johann (III) (1744–1807), Daniel (II) (1751–1834), and Jakob (II) (1759–1789). Johann (III), like his father, studied law but then turned to mathematics. When barely nineteen years old, he was called as a professor of mathematics to the Berlin Academy. He wrote on astronomy, the doctrine of chance, recurring decimals, and indeterminate equations. Jakob (II) also first studied law, but then became a professor of mathematics at the St. Petersburg Academy. His works are related to those of his uncle and teacher, Daniel Bernoulli.

Lesser Bernoulli descendants are Daniel (II), Christoph (1782–

1863), a son of Daniel (II), and Johann Gustav (1811–1863), a son of Christoph.

The Bernoulli relationships are summarized in the genealogical table of Figure 27.

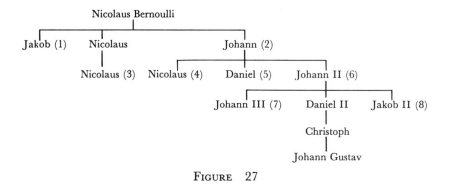

FIGURE 27

204° *A blow to primogeniture.* There are some who believe that the first-born child of a pair of parents inherits the greatest gifts from the parents. In connection with this belief it is of interest to point out that Jakob Bernoulli was the fifth child in his father's family, and Johann Bernoulli was the tenth.

205° *Eadem mutata resurgo.* Jakob Bernoulli was struck by the way the equiangular (or logarithmic) spiral $r = a^\theta$ reproduces itself in its involute, its evolute, and its caustics of both reflection and refraction. In imitation of Archimedes, he expressed a desire that such a spiral be engraved on his tombstone, along with the inscription *Eadem mutata resurgo* ("I shall arise the same, though changed")—a happy allusion to Christian hope. Visitors to the cloisters at Basel may still see the rough attempt of the stonecutter to carry out Jakob Bernoulli's desire.

206° *Invito patre sidera verso.* Like many fathers, Jakob Bernoulli's father tried to force an uncongenial vocation on his son. He wanted Jakob to become a theologian, and strenuously opposed the young man's interest in astronomy and mathematics. This led Jakob to adopt for his device Phaëthon driving the chariot of the sun, with the motto *Invito patre sidero verso* ("I study the stars against my father's will").

207° *Johann Bernoulli's nastiness.* Johann Bernoulli was thirteen years younger than his brother Jakob Bernoulli. It may be that a resulting attitude of superiority on Jakob's part led to Johann's deep jealous resentment. In any event, ill feeling long existed between the two brothers, and Johann often behaved very unfairly when others failed to recognize what he regarded as his just merits. Thus, in desperation, he once attempted to replace an incorrect solution of his own on the problem of isoperimetrical curves by another solution given by his brother Jakob. And he jealously expelled his son Daniel from his house for obtaining a prize from the French Academy that he himself had expected to receive.

208° *The origin of L'Hôpital's rule.* [The following is adapted, with permission, from the article of the same title by Professor D. J. Struik that appeared in the Historically Speaking section of *The Mathematics Teacher*, April, 1963, pp. 257–260.] The so-called rule of L'Hôpital, which states that

$$\lim_{x \to a} \frac{f(x)}{g(x)} = \frac{f'(a)}{g'(a)}$$

when $f(a) = g(a) = 0$, $g'(a) \neq 0$, was published for the first time by the French mathematician G. F. A. de l'Hôpital (or De Lhospital) in his *Analyse des infiniment petits* (Paris, 1696). The Marquis de l'Hôpital was an amateur mathematician who had become deeply interested in the new calculus presented to the learned world by Leibniz in two short papers, one in 1684 and the other in 1686. Not quite convinced that he could master the new and exciting branch of mathematics all by himself, L'Hôpital engaged, during some months of 1691–92, the services of the brilliant young Swiss physician and mathematician, Johann Bernoulli, first at his Paris home and later at his chateau in the country. When Bernoulli left for his home town Basel, the Marquis kept up correspondence with his tutor, at the same time publishing some original contributions of his own findings. When, in 1696,

L'Hôpital's book appeared, he acknowledged his indebtedness to Leibniz and Bernoulli, but only in general terms: "I have made free with their discoveries, so that whatever they please to claim as their own I frankly return to them."

The question of the actual dependence of L'Hôpital on Bernoulli remained unanswered, and acquired in the course of the years somewhat the character of a mystery. Bernoulli, after L'Hôpital had sent him a copy of the book, thanked him courteously and praised it. But subsequently, in some private letters written during the lifetime of the Marquis, he claimed that much of the content of the *Analyse des infiniment petits* was really his own property. In 1704, after L'Hôpital's death, he made a public claim to that section, No. 163, which contains the rule for 0/0. Mathematicians interested in this kind of priority puzzle have been philosophizing about this supposed dependence of L'Hôpital on Johann Bernoulli ever since, weighing Bernoulli's acknowledged greatness as a mathematician against his equally acknowledged reputation for nastiness. A generally acceptable conclusion was not reached until recent times.

Considerable clarification came in 1922, when Johann Bernoulli's manuscript on the differential calculus, dating from 1691–92, was at last published (the corresponding manuscript on the integral calculus was known from Johann Bernoulli's *Opera*, published in 1742, during the lifetime of the author). Comparison of these notes by Bernoulli and the text of L'Hôpital's book revealed that there was a considerable overlapping, so that it seemed that Bernoulli had fathered much of the nobleman's intellectual offspring. But the true situation came to light only in 1955, when Bernoulli's early correspondence was published. It then appeared that in 1694 a deal was actually made between the Marquis and his former tutor, by which L'Hôpital offered him a yearly allowance of 300 livres (and more later) provided that Bernoulli agreed to three conditions:

1. to work on all mathematical problems sent to him by the Marquis,

2. to make all the discoveries known to him, and

3. to abstain from passing on to others a copy of the notes sent to L'Hôpital.

This settled the priority question. Here is a translation of the section of the letter that contains the unusual proposition, sent by L'Hôpital in Paris to Johann Bernoulli in Basel, March 17, 1694:

> I shall give you with pleasure a pension of three hundred livres, which will begin on the first of January of the present year, and I shall send two hundred livres for the first half of the year because of the journals you have sent, and it will be one hundred and fifty livres for the other half of the year, and so on in the future. I promise to increase this pension soon, since I know it to be very moderate, and I shall do this as soon as my affairs are a little less confused. . . . I am not so unreasonable as to ask for this all your time, but I shall ask you to give me occasionally some hours of your time to work on what I shall ask you—and also to communicate to me your discoveries, with the request not to mention them to others. I also ask you to send neither to M. Varignon nor to others copies of the notes that you let me have, for it would not please me if they were made public. Send me your answer to all this and believe me, *Monsieur tout à vous*
>
> LE M. DE LHOSPITAL

Bernoulli's answer has not been found, but from a letter of July 22, 1694, we know that he had accepted the proposal. It must have been a little windfall for the impecunious young scientist, just married and still looking for a position (which he obtained the following year at the University of Groningen in the Netherlands). How long this interesting relationship lasted we do not know, but Bernoulli's finances improved and those of L'Hôpital did not become any better. By 1695 it may have come to an end.

Several letters from Bernoulli to his patron with answers to questions have now been published, and the one dated July 22, 1694, contains the rule for 0/0. The formulation is very much like the one we find in the *Analyse des infiniment petits*. Bernoulli's examples are also almost the same ones that L'Hôpital uses.

The situation has thus been clarified. When L'Hôpital's book appeared, Bernoulli was bound by his promise not to reveal which sections of the book belonged to him. He could only express himself privately. Then, after the death of the Marquis, he felt that he need not be so silent any more, and claimed as his own the most striking result in the book—the rule for 0/0. But he could not prove his assertion. At present he stands vindicated.

209° *Daniel Bernoulli's two little adventures.* Dr. Charles Hutton, in Volume I of his *A Philosophical and Mathematical Dictionary* (London, 1815), has recorded two little stories about Daniel Bernoulli, with the comment that Bernoulli claimed the involved adventures gave him more pleasure than all the honors that had fallen his way.

One day Daniel Bernoulli found himself traveling with a learned stranger who, taking delight in Bernoulli's conversation, asked him his name. "I am Daniel Bernoulli," he replied with attempted modesty. "And I," said the stranger, thinking Bernoulli meant to make sport of him, "am Isaac Newton."

On another occasion, Daniel Bernoulli invited the celebrated Swiss mathematician Samuel Koenig (1712–1757) to dine with him. During the dinner Koenig boasted, with some measure of self-pleasure, of a difficult problem that he had managed with much trouble to solve. Bernoulli continued doing the honors of his table, and when the two men retired to drink their coffee, he presented Koenig with a solution to the problem more elegant than Koenig's own.

210° *The apple of discord.* [The following is a reproduction, with permission, of the article entitled *Brachistochrone, tautochrone, cycloid—apple of discord*, by Professor J. P. Phillips of the University of Louisville, that appeared in the Historically Speaking section of *The Mathematics Teacher*, May, 1967, pp. 506–508.] The curve (see Figure 28) generated by the motion of a point on the circumference of a wheel

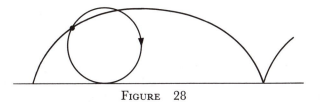

FIGURE 28

rolling on a flat surface is the cycloid, also known for its mechanical properties as the brachistochrone and the tautochrone, and also described by one mathematics historian as an apple of discord and by others as the Helen of geometry—beautiful, but the source of warfare.

From the moment of its invention, the cycloid generated controversy; indeed, the invention itself is not without dispute. Most sources, but not all, agree that it was not known to the Greeks. John Wallis in the seventeenth century credited it to Nicholas of Cusa in the fifteenth century, but he was mistaken. Certainly the curve is described in a book by Charles Bouvelles written in 1501, and he is usually considered the true inventor.

The greatest mathematicians of the Scientific Revolution of the seventeenth century were almost obsessed by the striking properties of the cycloid. The list of those who worked on it includes Galileo, Pascal, Roberval, Torricelli (who was said to have died of chagrin at intimations that he had plagiarized its quadrature), Descartes, Fermat, Wren, Wallis, Huygens, Johann Bernoulli, Leibniz, and Newton. This is very nearly the complete galaxy of the stars of the Scientific Revolution, and their common interest in the cycloid requires explanation. For that matter, their unceasing squabbles for priority as each new property was discovered constitute an interesting psychological question.

The intense interest in the seventeenth century in mechanics and the mathematics of motion accounts for much of the significance that the cycloid acquired at that time. Much of what was learned about it was produced by the methods of the calculus at a time when calculus had not yet been officially created, and some confusion as to methodology between individual mathematicians was thus inevitable.

Galileo was the first great man to take an interest in the cycloid, and, in what might be called an example of the true spirit of Archimedes, he determined experimentally about 1599 that its area is close to three times that of the generating circle, and went on to suggest that the arch of the cycloid ought to be suitable for building bridges. Such bridges were built later. Galileo also bequeathed an interest in cycloids to his students Torricelli and Viviani, as will be seen.

In the 1630's the French geometer Roberval offered a mathematical proof of the exact equivalence of the area of the cycloidal arch to three times that of the circle, and for his pains was attacked by Descartes for having labored overmuch to produce so small a result. In rebuttal to Descartes' simpler proof, Roberval commented that prior knowledge of the answer to be found had no doubt been of assistance.

Descartes met the insult by finding the tangent to the cycloid and challenged Roberval and Fermat to achieve this result for themselves; Fermat did, but Roberval did not or could not.

Since Roberval's results were not published at this time, he became embroiled several years later in further argument about priority of proof, this time with Torricelli, who had almost certainly achieved the area result independently. When Roberval accused him of plagiarism, Torricelli more or less promptly died, and doubtless over-romantic tradition has attributed his death to dismay at such an accusation. The tangent was again found independently by Viviani, another Italian pupil of Galileo.

Systematic study of the geometry of the cycloid began with Pascal. According to the story, he was suffering from toothache and assorted other pains one night and began to ruminate on the cycloid, at which point the pains abated. Although he had already withdrawn from mathematics to become a religious mystic, Pascal took this as an intimation that God approved of one last mathematical fling, and in eight days he worked out the area of the section produced by any line parallel to the base, the volumes generated by revolution about axis or base, the centers of gravity of these solids, and divers other results. To accomplish these results required the integration of several trigonometric functions, an operation never previously attempted and of course not recognized at the time (1658) as part of the calculus.

Pascal issued a challenge to the mathematicians of Europe to solve these problems also, but only two formal answers were received and neither was good enough to win the prize. One answer came from Wallis, who later rectified a number of errors in it and published it. Informally, Huygens, Wren, and Fermat solved parts of the challenge problem, and Huygens went on to new and different ground.

With Huygens, the primary interest in the cycloid shifts from its mathematical to its mechanical properties. As a purely mathematical accomplishment, Huygens discovered that the evolute of the cycloid is an equal cycloid (see Figure 29), but he appears to have been much more interested in applications of his discovery that the cycloid is the tautochrone; that is, a body freely accelerated by gravity along an inverted cycloidal arc requires the same length of time to reach the bottom of the arc no matter how far from the bottom it is to start

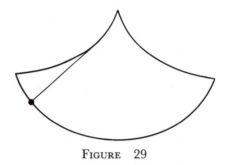

FIGURE　29

with. His proposal that pendulum clocks (which he invented) should use a cycloidal motion turned out to offer numerous mechanical difficulties, however.

The most famous discovery concerning the cycloid came in 1696, when Johann Bernoulli undertook to determine the equation of the path down which a particle will fall from one point to another in the shortest time. This was the first formal investigation of the calculus of variations, but is additionally important because Bernoulli saw an opportunity to use it in the controversy then raging between partisans of Newton and Leibniz concerning the invention of the calculus. In correspondence, Leibniz and Bernoulli agreed that only a handful of mathematicians in Europe besides themselves could possibly find the solution—the brachistochrone or cycloid—and that those who could would thereby demonstrate the most thorough mastery of the calculus. The problem was duly submitted as a challenge to mathematicians.

In a celebrated (but poorly documented) performance, Newton is said to have received the problem in the mail after a hard day's work at the Mint and to have solved it before the next morning. His solution was published anonymously in the *Philosophical Transactions of the Royal Society*, and Leibniz and Bernoulli were sorely disappointed.

Regrettably, the incident only stirred up Newton's abnormal suspicions that others were trying to claim credit for his work, and several years of underhanded efforts to undermine Leibniz's claim to the calculus followed. The whole episode became a disgrace to the principals and swayed the entire future of mathematics in Europe.

THE SMALL INITIAL UNDERSTANDING OF
THE CALCULUS

Of the many remarkable mathematical discoveries that were made in the seventeenth century, and that rendered that century outstandingly productive in the development of the subject, unquestionably the most remarkable was the invention of the calculus, toward the end of the century, by Isaac Newton and Gottfried Wilhelm Leibniz. The new tool proved to be almost unbelievably powerful in its astonishingly successful disposal of hosts of problems that had been baffling and quite unassailable in earlier days. Its general methods were able to cope with such matters as lengths of curved arcs, planar areas bounded by quite arbitrary curves, surface areas and volumes of all sorts of solids, intricate maximization and minimization problems, all kinds of problems involving related rates of change, geometrical questions about tangents and normals, asymptotes, envelopes, and curvature, and physical questions about work, energy, power, pressure, centers of gravity, inertia, and gravitational attraction. It was natural that this wide and amazing applicability of the new subject should attract mathematical researchers of the day, and that papers should be turned out in great profusion with seemingly little concern regarding the very unsatisfactory foundations of the subject. It was much more exciting to apply the marvelous new tool than to examine its logical soundness, for, after all, the processes employed justified themselves to the researchers in view of the fact that they worked.

211° *Newton's shift in the hypothesis.* Although for almost a hundred years after the invention of the calculus by Newton and Leibniz little serious work was done to strengthen logically the underpinning of its rapidly growing superstructure, it must not be supposed that there was no criticism of the existing weak base. Long controversies were carried on by some mathematicians, and even the two founders themselves were dissatisfied with their accounts of the fundamental concepts of the subject. One of the ablest criticisms of the faulty foundations came from a non-mathematician, the eminent metaphysician

Bishop George Berkeley (1685–1753), who insisted that the development of the calculus involved the logical fallacy of a shift in the hypothesis. We can clarify this particular criticism by considering one of Newton's approaches to what is now called *differentiation*.

In his *Quadrature of Curves* of 1704, Newton determines the derivative (or fluxion, as he called it) of x^3 as follows. We here paraphrase Newton's treatment:

> In the same time that x, by growing, becomes $x + o$, the power x^3 becomes $(x + o)^3$, or
>
> $$x^3 + 3x^2o + 3xo^2 + o^3,$$
>
> and the growths, or increments,
>
> $$o \quad \text{and} \quad 3x^2o + 3xo^2 + o^3$$
>
> are to each other as
>
> $$1 \quad \text{to} \quad 3x^2 + 3xo + o^2.$$
>
> Now let the increments vanish, and their last proportion will be 1 to $3x^2$, whence the rate of change of x^3 with respect to x is $3x^2$.

The shift of hypothesis to which Bishop Berkeley objected is evident; in part of the argument o is assumed to be nonvanishing, while in another part it is taken to be zero. Replies were made to Bishop Berkeley's criticism, but, without a logically rigorous treatment of limits, the objection could not be well met. Alternative approaches proved to be no less confusing. Indeed, all the early explanations of the processes of the calculus are obscure, encumbered with difficulties and objections, and not easy to read.

212° *Ghosts of departed quantities.*   In Section 35 of his publication *The Analyst*, Bishop Berkeley says the following about fluxions, or (as we call them today) derivatives: "And what are these fluxions? The velocities of evanescent increments. And what are these same evanescent increments? They are neither finite quantities, nor quantities infinitely small, nor yet nothing. May we not call them ghosts of departed quantities?"

213° *Address to an infidel mathematician.*   Bishop Berkeley's *The Analyst*, which proved to be a highly useful criticism of Newton's differential calculus, was originally designed as a reply to the "infidel mathematician" Edmund Halley (1656–1742), who professed to be an atheist. A friend of the Bishop, when in his sickbed, refused spiritual consolation because the great mathematician Halley had convinced him of the inconceivability of the doctrines of Christianity. This incident induced Bishop Berkeley to write *The Analyst*, to prove to the "infidel mathematician" that the principles of fluxions are no clearer than those of Christianity. In Section 7 of *The Analyst*, Berkeley says: "He who can digest a second or third fluxion, a second or third difference, need not, methinks, be squeamish about any point in Divinity."

A story is told that Halley once jested in Newton's presence about some theological questions, and Newton, who was deeply religious, repulsed him with the remark, "I have studied these things; you have not."

214° *Johann Bernoulli's postulate.* Some of the early explanations of the calculus border on the mystical and the absurd, as the statement by Johann Bernoulli that "a quantity which is increased or decreased by an infinitely small quantity is neither increased nor decreased."

215° *Guido Grandi's mysticism.* The whole matter of the foundations of the calculus was, for years, a subject of debate, as were all matters relating to infinite processes. This mystical period in the foundations of the calculus often itself invoked further mysticism that went considerably beyond that of the inventors of the subject. Thus the priest Luigi Guido Grandi (1671–1742), who was a professor first of philosophy and then of mathematics at the University of Pisa, and who wrote a number of works in geometry, considered the alternating series

$$1 - 1 + 1 - 1 + 1 - 1 + \cdots.$$

If we are to assign a value S to this sum, what should it be? If we group the terms of the series in one way, we obtain

$$S = (1 - 1) + (1 - 1) + (1 - 1) + \cdots$$
$$= 0 + 0 + 0 + \cdots = 0,$$

while if we group the terms another way we obtain

$$S = 1 - (1 - 1) - (1 - 1) - (1 - 1) - \cdots$$
$$= 1 - 0 - 0 - 0 - \cdots (= 1.$$

Faced with these two possibilities, some mathematicians argued that since 0 and 1 are equally probable sums, the correct sum of the series is the average value $\frac{1}{2}$. Of course, this value, too, can be obtained in a purely formal manner, for we have

$$S = 1 - (1 - 1 + 1 - 1 + 1 - 1 + \cdots) = 1 - S,$$

whence $2S = 1$, or $S = \frac{1}{2}$. Grandi was one of those who felt the sum of the series should be $\frac{1}{2}$. He supported this contention by citing the case of a father who bequeaths a gem to his two sons under the stipulation that each son is to keep the jewel every other year; it thus belongs to each son one half of the time. Grandi then went on and suggested that the formula

$$\frac{1}{2} = 1 - 1 + 1 - 1 + 1 - 1 + \cdots$$
$$= (1 - 1) + (1 - 1) + (1 - 1) + \cdots$$
$$= 0 + 0 + 0 + \cdots$$

indicates Creation from Nothing.

216° *Euler's formalism.* When the theory of a mathematical operation is only poorly understood, there is the danger that the operation will be applied in a blindly formal and perhaps illogical manner. The performer, not aware of possible limitations upon the operation, is likely to use the operation in instances where it does not necessarily apply. This is essentially what happened in analysis during the century following the invention of the calculus. Attracted by the powerful applicability of the subject, and lacking a real understanding of the foundations upon which the subject must rest, mathematicians manipulated analytical processes in an almost blind manner, often being guided only by a native intuition of what it was felt must be valid. A gradual accumulation of absurdities was bound to result, until, as a natural reaction to the pell-mell employment of intuition and formal manipulation, some conscientious mathematicians felt impelled to attempt the difficult task of establishing a rigorous foundation under the subject.

The work of the great Swiss mathematician Leonhard Euler (1707–1783) represents the outstanding example of eighteenth-century formal manipulation in analysis. It was by purely formal devices that Euler discovered the remarkable formula

$$e^{ix} = \cos x + i \sin x,$$

which, for $x = \pi$, yields

$$e^{i\pi} + 1 = 0,$$

a relation connecting five of the most important numbers of mathematics. By formal manipulations Euler arrived at an enormous number of curious relations like

$$i \log_e i = -(\pi/2),$$

and succeeded in showing that any nonzero real number n has an infinite number of logarithms (for a given base), all imaginary if $n < 0$ and all imaginary but one if $n > 0$. The beta and gamma functions of advanced calculus, and many other topics in analysis, similarly originated with Euler. He was a most prolific writer on mathematics and his name is attached to practically every branch of the subject.

It was Euler's remarkable mathematical intuition that generally held him on the right path, but there are, nevertheless, numerous instances where his formal manipulation led him into absurdities. For example, if the binomial theorem is applied formally to $(1 - 2)^{-1}$, we find

$$-1 = 1 + 2 + 4 + 8 + 16 + \cdots,$$

a result that caused Euler no wonderment! Also, by adding the two series

$$x + x^2 + \cdots = \frac{x}{1 - x} \quad \text{and} \quad 1 + \frac{1}{x} + \frac{1}{x^2} + \cdots = \frac{x}{x - 1},$$

Euler found that

$$\cdots + \frac{1}{x^2} + \frac{1}{x} + 1 + x + x^2 + \cdots = 0.$$

31

217° *Jean-le-Rond d'Alembert.* The first suggestion of a real remedy for the unsatisfactory state of the foundations of analysis came from Jean-le-Rond d'Alembert (1717–1783), who very correctly observed in 1754 that a theory of limits was needed. However, a sound development of this theory was not forthcoming until 1821, when the challenge was taken up and beautifully supplied by the great French mathematician Augustin-Louis Cauchy (1789–1857).

At the age of twenty-four, d'Alembert was admitted to the French Academy. In 1743 he published his *Traité de dynamique*, based upon the fruitful principle of kinetics that now bears his name. In 1744 he applied his principle in a treatise on the equilibrium and motion of fluids, and in 1746 in a treatise on the causes of winds. In each of these works, and also in one of 1747 devoted to vibrating strings, he was led to partial differential equations, and he became a pioneer in the study of such equations. With the aid of his principle he was able to obtain a complete solution of the baffling problem of the precession of the equinoxes. It was in 1754 that d'Alembert became permanent secretary of the French Academy. During his later years he worked on the great French *Encyclopédie*, which had been begun by Denis Diderot and himself.

But d'Alembert is one of those names in the history of mathematics to which is attached a special human interest apart from the recital of his scientific accomplishments. On the cold and blustery night of November 16, 1717, a gendarme, while making his rounds in Paris, heard the plaintive cries of an infant. Making his way toward the sound, he discovered a newly-born infant abandoned on the steps of the little Church of Saint Jean-le-Rond. The gendarme took the child to the parish commissary, who had it hurriedly christened Jean-le-Rond, after the name of its first resting place. Foster parents were found in a humble glazier and his wife who lived nearby and who gave the youngster love and a home. Later, for reasons not known, Jean-le-Rond added the name d'Alembert.

D'Alembert's mother was Madame de Tencin, sister of a cardinal. Her insatiable ambitions and tireless ministrations of intrigue found the illicit child a nuisance and a bother. The father, General Destouches, was a simple man of large heart who was easily bent by the mother's overbearing will. While the child grew up unrecognized by his mother, the father suffered searing pangs of conscience, and he tried in numerous ways to help the youngster. Thus he sent sums of money to the foster parents and, at his death nine years after the date of the abandonment, left enough to provide the boy with a good education.

In his life, d'Alembert exhibited considerably more intelligence than character. It is through the magnificence of the former that he secured immortality, and it is to the weakness of the latter that he owed the many sorrows and failures of his life. The moral is that, though intellect may win one a future world, it is by character that one wins the present world.

A famous and oft-quoted remark made by d'Alembert (and well worth citing on occasion in an elementary algebra class) is: "Algebra is generous; she often gives more than is asked of her."

BONAVENTURA CAVALIERI, YOSHIDA KŌYŪ, AND SEKI KŌWA

218° *Cavalieri's method of indivisibles.* Bonaventura Cavalieri was born in Milan in 1598, studied under Galileo, and was a professor of mathematics at the University of Bologna from 1629 until his death in 1647. He was one of the most influential mathematicians of his time and wrote a number of works on mathematics, optics, and astronomy. He was largely responsible for the early introduction of logarithms into Italy. But his greatest contribution was a treatise, published in its first form in 1635, devoted to the method of indivisibles—a method that probably can be traced back to Democritus.

Cavalieri's treatise is verbose and not clearly written, and it is difficult to know precisely what is to be understood by an "indivisible." It seems that an indivisible of a given planar piece is a chord of the piece, and an indivisible of a given solid is a planar section of that solid. A planar piece is considered as made up of an infinite set of parallel chords and a solid as made up of an infinite set of parallel plane sections. Now, Cavalieri argued, if we slide each member of the set of parallel chords of some planar piece along its own axis, so that the end points of the chords still trace a continuous boundary, then the area of the new planar piece so formed is the same as that of the original planar piece. A similar sliding of the plane sections of a given solid yields another solid having the same volume as the original one. These

33

results give the so-called *Cavalieri's principle*: (1) *If two planar pieces are included between a pair of parallel lines, and if the two segments cut by them on any line parallel to the including lines are equal in length, then the areas of the planar pieces are equal;* (2) *if two solids are included between a pair of parallel planes, and if the two sections cut by them on any plane parallel to the including planes are equal in area, then the volumes of the solids are equal.*

Cavalieri's principle constitutes a valuable tool in the computation of areas and volumes and can easily be rigorously established. As an illustration of the principle, consider the following application leading to the formula for the volume of a sphere. In Figure 30 we have a

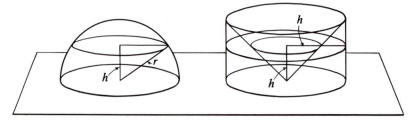

FIGURE   30

hemisphere of radius r on the left, and on the right a cylinder of radius r and altitude r with a cone removed whose base is the upper base of the cylinder and whose vertex is the center of the lower base of the cylinder. The hemisphere and the gouged-out cylinder are resting on a common plane. We now cut both solids by a plane parallel to the base plane and at distance h from it. This plane cuts the one solid in a circular section and the other in an annular, or ring-shaped, section. By elementary geometry we easily show that each of the two sections has an area equal to $\pi(r^2 - h^2)$. It follows, by Cavalieri's principle, that the two solids have equal volumes. Therefore the volume V of a sphere is given by

$$V = 2(\text{vol. of cylinder} - \text{vol. of cone})$$

$$= 2(\pi r^3 - \pi r^3/3) = 4\pi r^3/3.$$

The assumption and consistent use of Cavalieri's principle simplify the derivation of many formulas encountered in a high school

34

course in solid geometry. This procedure has been adopted by a number of textbook writers and has been advocated on pedagogical grounds.

219° *Some applications of Cavalieri's principle.* The interested reader may care to play with the following problems.

1. Devise a polyhedron that can be used as a comparison solid for obtaining the volume of a sphere of radius r by means of Cavalieri's principle. [Let AB and CD be two line segments in space such that (1) $AB = CD = 2r\sqrt{\pi}$, (2) AB and CD are each perpendicular to the line joining their midpoints, (3) AB is perpendicular to CD. The tetrahedron $ABCD$ may serve as the comparison polyhedron.]

2. An oblique plane through the center of the base of a right circular cylinder cuts off from the cylinder a cylindrical wedge called a *hoof*. Find, by Cavalieri's principle, the volume of a hoof in terms of the radius r of the associated cylinder and the altitude h of the hoof. [Divide the hoof into two equal parts by a plane m through the axis of the cylinder and let A be the area of the resulting triangular cross section of the hoof. Construct a prism having as its base a square of area A, the base lying in the plane m, and having an altitude equal to the radius r of the cylinder. Cut from this prism a pyramid whose base is the base of the prism not lying in m, and whose vertex is a point in the other base of the prism. The volume of the hoof is $2hr^2/3$.]

3. Find, by Cavalieri's principle, the volume of the *spherical ring* obtained by removing from a solid sphere a cylindrical boring that is coaxial with the polar axis of the sphere. [Use for a comparison solid a sphere with diameter equal to the altitude h of the ring.]

4. Show that all spherical rings of the same altitude have the same volume, irrespective of the radii of the spheres of the rings.

220° *An " a" for an " i."* Written histories often contain hidden perpetuated errors. Some historian will record an erroneous and undocumented statement, and then subsequent historians, leaning on the earlier work, will repeat the erroneous statement. Many such erroneous statements have been widely perpetuated over considerable periods of time.

An instance of the perpetuation of an error in the history of mathematics is a statement concerning the Italian mathematician

Bonaventura Cavalieri (1598–1647), whose method of indivisibles was a forerunner of the integral calculus. In a number of major encyclopedias, histories, and source books, it is stated that Cavalieri was a Jesuit. Now the truth of the matter is that Cavalieri was not a Jesuit, but a *Jesuat*.

The religious group known as Jesuats, and which was in no way related to the Jesuits, was founded in the fourteenth century by the Blessed John Colombini of Siena, the order being approved by Pope Urban V in 1367. The original work of the order was the care of the sick—especially of those stricken by the Black Death, which raged over Europe at the time—and the burial of the dead. With the passage of time the Jesuat order diminished, and in 1606 an attempt at a revival was made. It was a few years later that Cavalieri, when he was fifteen, was accepted as a member of the order.

Certain abuses later made their way into the Jesuat order, with the result that the group no longer exists today. It seems that the manufacture and sale of distilled liquors, apparently in a manner not acceptable to Canon Law, along with a growing scarcity of members, led to the order's suppression by Pope Clement IX in 1668.

221° *A synonym for "arithmetic."*   One of the early Japanese mathematicians of the seventeenth century was Yoshida Kōyū, or Mitsuyoshi (1598–1672), whose *Jinkō-ki* was the first great work on arithmetic to appear in Japan. The title of the work means "Small number, large number, treatise." That is, it was a treatise on numbers of all sizes. The interesting thing here is that the name of this work became so familiar in Japan that it was frequently used as a synonym for "arithmetic." One is reminded of the word *algorismus* (corruption of Al-Khowarizmi) as an early European synonym for "arithmetic," and of the name *Euclid* as a synonym for "geometry."

222° *The Japanese Newton.*   In the seventeenth century, Japan awoke to her intellectual possibilities, and it is interesting that her progress in mathematics had some similarities to that which was taking place at essentially the same time in Europe. Thus, around 1660–1670, Japan independently developed an approach to the integral calculus somewhat after the method of Cavalieri, and then later the *yenri*, or circle principle, which led to a kind of native calculus having the same aims as the invention of Newton and Leibniz.

The most outstanding Japanese mathematician of the seventeenth century was Seki Kōwa (1642–1708), who was born of a samurai family and showed strong mathematical talent at an early age. Seki Kōwa has, with some justification, been called the Japanese Newton. He was born in the same year as was Newton. Like Newton, he acquired much of his knowledge by self-study, was skilled in mechanics as well as mathematics, and was a brilliant problem solver. But the greatest similarity between the two men lies in the claim that Seki Kōwa was the creator of the *yenri*, or circle principle. Thus Seki Kōwa probably invented the calculus of the East just as Newton invented the calculus of the West. Both of these calculi were to receive considerable improvement and development by subsequent mathematicians. Newton was honored by royalty during his lifetime, when he was knighted; Seki Kōwa received his royal honor posthumously in 1907, when the Emperor of Japan bestowed upon him the highest such honor awarded a Japanese scholar. It is of further interest to note that Seki Kōwa anticipated Leibniz by about ten years in the discovery of determinants in connection with systems of simultaneous linear equations.

223° *Seki Kōwa as a prodigy.* There are stories that have come down to us concerning the early age at which Seki Kōwa showed mathematical ability. One story says that he was only five when he pointed out an arithmetical error made in his presence by his elders, who were so struck by his marvelous arithmetical abilities that they gave him the title of "divine child." This title was placed on a monument to Seki erected in Tokyo in 1794.

Another story says that when he was nine years old, Seki Kōwa saw a servant studying the *Jinkō-ki* of Yoshida Kōyū. When the servant became perplexed over a certain problem in the text, Seki showed him the proper solution.

224° *Seki Kōwa and his employer.* In later life, Seki Kōwa served in a public capacity as examiner of accounts to the Lord of Kōshū, just as Newton became Master of the Mint under Queen Anne. Some anecdotes are told about Seki Kōwa when working for the Lord of Kōshū. One of these tells that at one time, when on a mission for his Lord, Seki traveled in a palanquin and amused himself by noting the directions, distances, objects, elevations, and depressions along the way. From these observations he prepared a map of the journey that was so detailed and accurate that his employer exclaimed that "though he traveled like a samurai he observed like a geographer."

Another anecdote tells that his employer decided to distribute among members of his family equal portions of a large piece of precious incense wood. When the division was attempted, no one could find a way of meeting the demand that the portions be equal. Finally, appeal was made to Seki, who easily and cleverly accomplished the difficult task.

A third anecdote tells that the Emperor of China sent a wonderful clock to the Lord of Kōshū. The clock was so constructed that the figure of a man struck the hours. After some years the mechanism of the clock failed and the figure would no longer strike the bell. The most skillful artisans of the land were called in to repair the clock, but to no avail. When Seki heard of the trouble he asked if he might take the clock home and examine it. Soon he returned the clock to his Lord in perfect working condition.

SOME LESSER SEVENTEENTH- AND EIGHTEENTH-CENTURY BRITISH MATHEMATICIANS

225° *Up the ladder of success.* Edmund Gunter (1581–1626), one of Henry Briggs's colleagues, published in 1620 a seven-place table of the common logarithms of the sines and tangents of angles for intervals of a minute of arc. Gunter also constructed in 1620 a logarithmic scale, or a line of numbers on which the distances are proportional to the logarithms of the numbers indicated, and mechanically performed multiplications and divisions by adding and subtracting segments of this scale with the aid of a pair of dividers. In this way Gunter preluded the familiar slide rule of today. It was Gunter who invented the words *cosine* and *cotangent,* and he is known to surveyors for his "Gunter chain."

An amusing story is told about Gunter as a theological student at Christ Church. Apparently he was a very poor preacher, for when he was chosen to preach the Passion sermon, some divines who heard him declared that the Saviour never suffered so much since his Passion as in Gunter's sermon, it was such a lamentable effort. Gunter left the ministry to become, in 1619, professor of astronomy at Gresham College in London, apparently to the benefit of both his vocations.

226° *Halley's magnanimity and perseverance.* Edmund Halley (1656–1742) showed remarkable promise and progress in his early youth. When he was not yet twenty-one he set out for St. Helena to make astronomical observations in the neglected Southern Hemisphere. While at St. Helena he charted some three hundred fixed stars and made the first complete observation of a transit of Mercury, with the result that Flamsteed, the first Astronomer Royal of England, christened him the "southern Tycho." Halley was elected a member of the Royal Society when he was only twenty-two. Although his materialistic views prevented his appointment as Savilian Professor of Astronomy at Oxford in 1691, he did follow Wallis as Savilian Professor of Geometry in 1703, and he succeeded Flamsteed as Astronomer Royal in 1721. He was the first to predict the time of return of a specified comet, and the comet was accordingly named after him.

Halley was both magnanimous and persevering. The former trait is perhaps best illustrated by his relationship with Isaac Newton. It was only at Halley's insistence and with his personal financial backing that the *Principia* of Newton was published. Halley consistently exhibited kindness and generosity towards scholars.

Halley's perseverance is strikingly illustrated by his incredible translation from the Arabic of a work by Apollonius. This work had been thought lost until Edward Bernard, who was Savilian Professor of Astronomy from 1673 to 1691, discovered an Arabic copy of the work in the Bodleian Library. Being skilled in Oriental languages, Bernard started to translate the work from Arabic into Latin, but at his death he had scarcely completed a tenth of it. Halley, who had an interest in the ancient mathematical writers, undertook to finish the task. Undismayed by the fact that he did not know a word of Arabic, he carefully studied the thirteen pages completed by Bernard, first picking out those words whose meanings he could recognize from the context. He then examined the argument and turned over and over in his mind the possible meanings of the words he did not recognize. By this method of deciphering, he groped his way through the work to obtain a general idea of its content. Then, by going over the manuscript again and again, he managed to complete the translation with assistance from no one.

227° *"According to Cocker."*   There were a number of successful writers of elementary mathematical textbooks in the seventeenth century. Perhaps the most popular of the British textbook writers was Edward Cocker (1631–1675), who made a considerable name for himself as a teacher of writing and mathematics. He wrote several books on arithmetic and penmanship, but it was his *Arithmetick, being a Plain and Easy Method*, edited by John Hawkins in 1678, that introduced the expression "according to Cocker" into the common speech of England. The book ran through close to a hundred editions and had great influence upon British textbooks for over a hundred years.

The English expression "according to Cocker" reminds one of the similar German expression "nach Adam Riese" (see Item 127°).

228° *Man of promise.*   There have been a number of mathematicians, like Pascal, Galois, and Clifford, who died at relatively early ages and who showed remarkable promise in their field. One of these was Roger Cotes (1682–1716). Cotes had shown a penchant for mathematics when he was only twelve years old. When only twenty-four, he was appointed to the Plumian Professorship of Astronomy, as the first to fill this chair that had been established in 1704 by Dr. Plume, the Archdeacon of Rochester. In 1713, Cotes published at Cambridge the second edition of Newton's *Principia*. His mathematical writings were collected and published shortly after his untimely death. Newton once said, "If Cotes had lived, we had known something."

229° *Death in an arithmetic progression.*   Abraham de Moivre (1667–1754), though born in France, lived most of his life in London, and became an intimate friend of Isaac Newton. He is particularly noted for his work *Annuities upon Lives*, which played an important role in the history of actuarial mathematics, his *Doctrine of Chances*, which contained much new material on the theory of probability, and his *Miscellanea analytica*, which contributed to recurrent series, probability, and analytical trigonometry. He is credited with the first treatment of the probability integral and of (essentially) the normal frequency curve which is so important in statistics. The misnamed *Stirling's formula*, which says that for very large n

$$n! \approx \sqrt{2\pi n}\, n^n/e^n,$$

is due to De Moivre and is highly useful for approximating factorials of large numbers. The familiar formula

$$(\cos x + i \sin x)^n = \cos nx + i \sin nx, \qquad i = \sqrt{-1},$$

known by De Moivre's name but already stated in substance by Cotes, has become the keystone of analytic trigonometry.

De Moivre was forced to support himself by private tutoring and lecturing. It is said that he selected a London coffee house as his headquarters, and there made a reputation and a modest income by solving problems brought to him.

Interesting is the fable often told of De Moivre's death. According to the story, De Moivre once noticed that each day he required a quarter of an hour more sleep than on the preceding day. When the arithmetic progression reached twenty-four hours, De Moivre passed away.

230° *An inspiring achievement.* There is something thrilling, astonishing, inspiring, and also deflating, in the account of the life of a prodigy, and mathematics has had its share of prodigies. One of these was the Scotch mathematician Colin Maclaurin (1698–1746). Maclaurin matriculated at the University of Glasgow at the age of eleven and soon showed a liking for mathematics. When he was twelve he accidentally ran across a copy of Euclid's *Elements* and mastered the first six books in a few days. At fifteen he took his master's degree and gave a remarkable public defense of his thesis on the power of gravity. At nineteen he was elected to the chair of mathematics at the Marischal College in Aberdeen, and at twenty-one he published his first important work, a treatise on geometry, some of the propositions of which he had worked out when he was sixteen. He then traveled awhile as the tutor to the son of Lord Polwarthy. At twenty-seven he became the deputy, or assistant, to the professor of mathematics at the University of

Edinburgh. There was some difficulty in obtaining a salary to cover the assistantship, and Newton offered to bear the cost personally so that the university could secure the services of so outstanding a young man as Maclaurin. In time Maclaurin succeeded the man he assisted. His treatise on fluxions appeared when he was forty-four, only two years before he died; this was the first logical and systematic exposition of Newton's method of fluxions and was written by Maclaurin as a reply to Bishop Berkeley's attack on the principles of the calculus.

Maclaurin played an active role in the opposition to the Young Pretender in 1745. When the Highlanders approached Edinburgh, Maclaurin fled to York, but the exposure in the trenches at Edinburgh and the privations he had to suffer in his escape proved fatal to him.

231° *An inspiring character.* There are some men who inspire admiration because of some personal fortitude above and beyond their actual achievements. Such a man in mathematics was Nicholas Saunderson (1682–1739). Let us first list his principal professional accomplishments. He became Lucasian Professor of Mathematics at Cambridge in 1711, having first served as a deputy there. He was created doctor of laws in 1728 at the recommendation of George II. He was elected a Fellow of the Royal Society in 1736. He was outstandingly successful as a teacher and is particularly known for his *Algebra*, which was published posthumously in 1740–41. His *Method of Fluxions* was also published after his death, in 1751. He was personally admired by Newton, Cotes, De Moivre, and many other scholars, and he did much to make Newton's methods known to the mathematicians of his time.

A fairly impressive professional life, by any standards! How much the more remarkable, then, when we learn that when only one year old Saunderson became totally and permanently blind from an attack of smallpox. As the blind mathematician of Cambridge he won the admiration and respect of all who knew him. He developed the ability to carry on long and complicated mathematical arguments mentally.

232° *A noble epitaph.* There is a noble epitaph to Sir Christopher Wren (1632–1723), the mathematician who turned architect (see Item 150°). It appears in St. Paul's Cathedral, which Wren redesigned and rebuilt after the great London fire of 1666. When Wren died in 1723, he was buried in St. Paul's with the fitting epitaph, *Si monumentum requiris, circumspice* ("If you seek a monument, look about you").

233° *A hidden name.* John Flamsteed (1646–1719), the first Astronomer Royal of England, who was succeeded by Edmund Halley (see Item 226°), once wrote: "I crave the liberty to conceal my name, not to suppress it. I have composed the letters of it written in Latin in this sentence—*In Mathesi a sole fundes.*" Flamsteed's name in Latin is *Johannes Flamsteedius.*

SOME LESSER SEVENTEENTH- AND EIGHTEENTH-CENTURY CONTINENTAL MATHEMATICIANS

234° *The Van Schooten family.* After the manner of the Bernoulli family of Switzerland, the Van Schooten family of Holland produced three generations of professors of mathematics, but, unlike the Bernoullis, none of these men was of first rank. They were all professors at the University of Leyden, and they were all sympathetic to mathematics. The first was Frans van Schooten (1581–1646), who constructed a trigonometric table in 1627. The second, son of the first, was also named Frans van Schooten (1615–1660). He instructed engineering at Leyden and was the teacher of Christiaan Huygens. He edited Viète, wrote on perspective, and is best known for his 1649 Latin translation of Descartes' *Géométrie.* His half brother, Petrus van Schooten (1634–1679), for a time occupied the chair of mathematics at Leyden and then was transferred to the chair of Latin.

235° *The Clairaut family.* Another mathematically gifted family was the Clairaut family of France. Claude Alexis Clairaut was born in Paris in 1713 and died there in 1765. He was a youthful mathematical prodigy, composing in his eleventh year a treatise on curves of the third order. This early paper, and a singularly elegant subsequent one on the differential geometry of twisted curves in space, won him a seat in the French Academy of Sciences at the illegal age of eighteen. In 1736 he accompanied Maupertius on an expedition to Lapland to measure the length of a degree of one of the earth's meridians. In 1743, after his return to France, Clairaut published his definitive work,

Théorie de la figure de la Terre. In 1752 he won a prize from the St. Petersburg Academy for his paper *Théorie de la Lune*, a mathematical study of lunar motion that cleared up some to then unanswered questions. In 1759 he calculated, with an error of about a month, the 1759 return of Halley's comet. A scientific rivalry, often not friendly, existed between Clairaut and d'Alembert.

Clairaut had a brother who tragically died of smallpox when only sixteen, but who at fourteen read a paper on geometry before the French Academy and at fifteen published a work on geometry. The father of the Clairaut children was a teacher of mathematics, a correspondent of the Berlin Academy, and a writer on geometry.

These two Clairaut brothers, who must be regarded as two of the most precocious mathematicians of all times, were two of a family of twenty children only one of whom survived the father.

236° *The "great flattener."*   In 1736 Pierre Louis Moreau de Maupertius (1698–1759) led an expedition to Lapland to determine the length there of a degree of one of the earth's meridians. The expedition was undertaken to settle a dispute as to the shape of the earth. Newton and Huygens had concluded, from mathematical theory, that the earth is flattened at the poles, a contradiction of the Cartesian contention that the earth is elongated at the poles. Between 1700 and 1720, the Italian astronomer and mathematician Giovanni Domenico Cassini (1625–1712), and his French-born son Jacques Cassini (1677–1756), measured an arc of longitude in France extending from Dunkirk to Perpignan, and their result seemed to support the Cartesians. A controversy arose in which many mathematicians became involved. The outcome was that an expedition was sent in 1735 to Peru, followed the next year by Maupertius's expedition to Lapland. The measurements made in Peru and in Lapland unquestionably confirmed the Newton–Huygens belief, and earned Maupertius the title of *grand aplatisseur* ("great flattener"). Maupertius was rewarded by being named president of the Berlin Academy and for many years basked in the glory of his fame at the court of Frederick the Great.

The description of Maupertius as the man who flattened the earth reminds one of the description of Copernicus as the man who told the sun to stand still.

237° *A fall from the cold heights.* Maupertius's respectable position as president of the physical class of the Berlin Academy continued pleasantly until 1750, when he became embroiled in a heated controversy with the Swiss mathematician Samuel König concerning the discovery of the principle of least action in mechanics. Maupertius was seeking, as did Einstein in more recent years, a general principle which would subsume and unify the laws of the universe. He announced a cloudy formulation of the principle of least action and derived from it a "proof" of the existence of God. König accused Maupertius of plagiarism, and many ridiculed the excursion into theology. The controversy was brought to a climax when Voltaire wrote his *Diatribe du Dr. Akakis* (1752), to defend König and lampoon Maupertius for his metaphysical effort. Neither King Frederick's support nor Leonhard Euler's defense could save the "pompous and touchy" president, and the unhappy and deflated mathematician died a few years later in Basel at the home of the Bernoullis.

238° *The death of de Lagny.* Another lesser French mathematician of the period under consideration was Thomas-Fantel de Lagny (1660–1734), who wrote on methods of extracting roots, the construction of a cube having the same volume as that of a given sphere, binary arithmetic, and methods in problem solving. A curious story is told of his death. Maupertius was called to de Lagny's deathbed and, finding the poor man unconscious, suddenly asked him for the square of 12. Like an automaton, de Lagny rose in bed, gave the answer, and immediately passed away.

239° *Probabilité des jugements.* In the last half of the eighteenth century, efforts were made to apply the doctrine of chances to new fields. One of these attempts concerned an application to man's judgment, such as computing the probability that a tribunal will arrive at a fair verdict if each of the witnesses and jurymen has a number assigned to him measuring the probability that he will speak or understand the truth. This *probabilité des jugements*, with its overtones of Enlightenment philosophy, was prominent in the work of Antoine-Nicolas Caritat, Marquis de Condorcet (1743–1794). Among Condorcet's conclusions was that capital punishment should be abolished, because,

45

however great the probability of the correctness of a single decision, there will be a large probability that in the course of many decisions some innocent person will be wrongfully condemned.

Condorcet was admitted to the Académie des Sciences when only twenty-six, and at thirty he became the group's secretary. Though his ability in mathematics was quite appreciable, he is perhaps more noted for his studies in philosophy, literature, and politics. In his vain effort to guide the revolutionary torrent into a constitutional channel, he fell a victim to the republican terrorists. On March 28, 1794, he was arrested and jailed for the night somewhere near Bourg-la-Reine, the intention being to remove him to Paris the next day. But Condorcet had sworn that he would never be taken to Paris, and the next morning when his jailer opened the door of the cell, Condorcet was found dead, with an empty poison ring beside him. He had kept his promise.*

The manner of Condorcet's apprehension is interesting. Hiding out in disguise, he entered an eating house and unthinkingly ordered an aristocrat's omelet containing a dozen eggs. The surprised cook asked him his trade. "Carpenter," he replied. "Really? Let me see your hands. You are no carpenter." Caught in his lie, he was turned over under suspicion to the authorities and thus discovered to be the sought Marquis de Condorcet.

LEONHARD EULER

LEONHARD Euler was born in Basel, Switzerland, in 1707, and he studied mathematics there under Johann Bernoulli. In 1727 he acquired the chair of mathematics at the new St. Petersburg Academy formed by Peter the Great. Fourteen years later he accepted the invitation of Frederick the Great to go to Berlin to head the Prussian Academy. After twenty-five years in this post Euler returned to St. Petersburg, remaining there until his death in 1783 when he was seventy-six years old.

---

* Some believe he may have died of exhaustion from the exposure preceding his capture; others believe he may have died of apoplexy. Condorcet had welcomed the revolution and had done much to assist it. He was one of the first to declare for a republic, but he could not sanction the brutalities of the extremists.

Euler was a voluminous writer on mathematics, indeed, the most prolific writer in the history of the subject. His name is attached to every branch of the study. It is of interest to note that his amazing productivity was not in the least impaired when, about 1766, he had the misfortune to become totally blind.

In addition to being a great research writer for some of the journals of his time, Euler was also a masterful writer of textbooks, in which he presented his material with great clarity, detail, and completeness. These texts enjoyed a marked and a long popularity, and to this day make very interesting and profitable reading. One cannot but be surprised at Euler's enormous fertility of ideas, and it is no wonder that so many of the great mathematicians coming after him have admitted their indebtedness to him.

Euler was helped and encouraged by his father in the choice of mathematics as a vocation. His father was a Calvinist pastor with an interest in mathematics, having studied the subject under Jakob Bernoulli. The religious training of Euler's childhood remained with him throughout his life. His simple, unquestioning faith enabled him to accept blindness with courage, and also made it difficult for him to harmonize with such men as Voltaire and Frederick the Great. Euler married twice and had a total of thirteen children, all but five of whom died when young. His first son, Johann Albrecht Euler (1734–1800), attained some fame in the field of physics.

240° *Tributes to Euler.* Many glowing tributes have been paid to Euler, such as the following three, the first two of which were made by F. Arago, physicist and astronomer, and the third one by F. Rudio, historian of mathematics.

"Euler could have been called, almost without metaphor and certainly without hyperbole, analysis incarnate."

"Euler calculated without any apparent effort, just as men breathe and as eagles sustain themselves in the air."

"We may safely say that the whole form of modern mathematical thinking was created by Euler. It is only with the greatest difficulty that one is able to follow the writings of any author immediately preceding Euler, because it was not yet known how to let the formulas speak for themselves. This art Euler was the first one to teach."

241° *Euler's blindness.*   In 1735, the year after his wedding and when he was in Russia, Euler received a problem in celestial mechanics from the French Academy. Though other mathematicians had required several months to solve this problem, Euler, using improved methods of his own and by devoting intense concentration to it, solved it in three days and the better part of the two intervening nights. (Later, with still superior methods, Gauss solved the same problem within an hour!) The strain of the effort induced a fever from which Euler finally recovered, but with the loss of the sight of his right eye. Stoically accepting the misfortune, he commented, "Now I will have less distraction."

Thirty-one years later, in 1766, when he was again in Russia, Euler developed a cataract in his remaining eye and went completely blind. Now blindness would seem to be an insurmountable barrier to a mathematician, but, like Beethoven's loss of hearing, Euler's loss of sight in no way impaired his amazing productivity. He continued his creative work by dictating to a secretary and by writing formulas in chalk on a large slate for his secretary to copy down.

In 1771, after five years in darkness, Euler underwent an operation to remove the cataract from his left eye, and for a brief period he was able to see again. But within a few weeks a very painful infection set in, and when it was over Euler was once again totally blind—so to continue for the remaining twelve years of his life.

242° *Euler's memory.*   Euler possessed a phenomenal memory. He seems to have been one of those very rare people who almost perfectly retain in their minds anything that they have once read or heard. This great memory served Euler in good stead during his years of blindness. He calculated long and difficult problems on the blackboard of his mind, sometimes carrying out arithmetical operations to over fifty decimal places. It is said that he once memorized the *Aeneid* and was ever after able to recite the entire Latin work word for word, remarking where each page of his copy ended and the next one began.

243° *Euler's concentration.*   Most mathematicians require the absolutely undisturbed peace and quiet of solitude when engaged in creative work. Euler, whose mind worked at lightning speed, was capable of such intense concentration that he needed neither quiet nor solitude. Indeed, most of his creative work was done at home amidst the bedlam and clamor caused by several small children playing noisily around his desk. The great mathematician would remain serene and would often rock a baby with one hand while working out a difficult problem with the other. He permitted himself to be repeatedly interrupted, and after each interruption he would continue from where he had left off without any loss of either temper or train of thought.

244° *Euler's productivity.*   Without question, Euler was the most prolific writer of mathematics in the entire history of the subject. He wrote mathematics incessantly with the effortless ease and fluency that a skilled conversationalist exhibits when speaking. It is said that he would dash off a mathematical paper in the scant half-hour between the first and second call to dinner.

As soon as Euler finished a paper, he would put it on top of a growing stack that was awaiting the printer. When material was needed to fill the Academy's transactions, the printer would grab a sheaf from the top of the pile. In this way it often happened that the dates of publication of Euler's papers ran counter to those of their composition. Particular confusion resulted when this happened to a progressive sequence of papers on the same subject.

Euler promised to provide the Academy with papers sufficient to enrich its transactions for twenty years. This promise was more than fulfilled, for, though Euler died in 1783, each volume of the transactions down to 1818 usually contained at least one of his papers.

A monumental edition of Euler's complete works was initiated in 1909 by the Swiss Society of Natural Sciences and was planned to run into seventy-three large quarto volumes, but the later discovery of overlooked material will considerably increase this number. Wars and subsequent inflations drastically disrupted the original endowment, but it is hoped that world-wide subscription by individual mathematicians and mathematical societies will see it through. It is only proper that the world at large bear the brunt of the expenses, for Euler belongs to the entire civilized globe and not only to Switzerland.

It is natural to wonder who has the honor of being the second most prolific writer of mathematics. It appears that this honor should go either to the English mathematician Arthur Cayley (1821–1895)—whom not many would have guessed—or to the French mathematician Augustin-Louis Cauchy (1789–1857). An actual counting of the pages of production of these two men may be needed to determine which one holds the honor.

245° *Euler's universality.*   Euler's knowledge and interest were by no means confined to just mathematics and physics. He was an excellent general scholar, with especially extensive knowledge of astronomy, medicine, botany, and chemistry. He attentively read the eminent Roman writers, was well informed on both the civil and the literary history of all ages and nations, and showed a wide acquaintance with many branches of literature. Undoubtedly he was greatly aided in these diverse fields by his uncommon memory.

When Euler's two friends, Daniel and Nicholas Bernoulli, joined the Petersburg Academy, they wrote back to Euler in Switzerland of an opening in the Academy in the section of physiology and urged Euler to brush up on the subject and apply for the position. This Euler did, and was accepted at the Petersburg Academy at the age of twenty. Fortunately, because of a disorganization at the Academy brought on by the death of Catherine I, who died on the very day Euler arrived in St. Petersburg, the medical position was overlooked and Euler unexpectedly slipped into a position in the mathematics section.

246° *The origin of a paper in mechanics.*   Euler had a great appreciation for Virgil's *Aeneid*. Not only could he recite the entire Latin poem word for word from beginning to end, but there is a paper of his which, as he himself has informed us, owed its origin to a verse of the poem—the line that reads (in translation), "The anchor drops, the rushing keel is staid."

Every now and then a mathematical problem originates in literature. One recalls, for example, the famous duplication-of-the-cube problem of antiquity, wherein one is to construct with straightedge and compasses the edge of a cube having exactly twice the volume of a given cube. There is evidence that this problem may have originated in the words of some mathematically unschooled and obscure ancient Greek poet who represented the mythical King Minos as dissatisfied with the size of a tomb erected to his son Glaucus. Minos ordered that the tomb be doubled in size. The poet then had Minos add, incorrectly, that this can be accomplished by doubling each dimension of the tomb. This faulty mathematics on the part of the poet led the geometers to take up the problem of finding how one can double a given solid while keeping the same shape.

Again, only a few years ago, a problem in the Problem Department of *The American Mathematical Monthly* originated in the following lines of *Mrs. Miniver*:

> She saw every relationship as a pair of intersecting circles. The more they overlapped, it would seem at first glance, the better the relationship; but this is not so. Beyond a certain point, the law of diminishing returns sets in, and there are not enough private resources left on either side to enrich the life that is shared. Probably perfection is reached when the area of the two outer crescents, added together, is exactly equal to that of the leaf-shaped piece in the middle. On paper there must be some neat mathematical formula for arriving at this; in life, none.

The problem was to discuss the possibility of a unique solution for circles of given radii.

247° *Euler, the supreme calculator.*   Condorcet (see Item 239°), in his fine *Éloge de M. Euler*, pronounced before the Académie des Sciences in 1785, tells of two of Euler's students who had tediously and independently summed to seventeen terms a complicated convergent series, only to find disagreement in the fiftieth decimal places of their results. In an appeal to the master to determine which answer was correct, Euler performed the whole calculation *mentally* and soon settled the dispute.

248° *Euler joins the Prussian Academy.*   When thirty-five, Euler was invited by Frederick the Great to come to Berlin and adorn the Prussian Academy. Euler readily accepted, happy to leave a Russia that Empress Anna had bathed in blood in her relentless effort to clean out spies and traitors. The Prussian Dowager Queen took a fancy to Euler and tried unsuccessfully to make him more communicative. Upon inquiring why he was so reticent, Euler replied, "Madame, I come from a country where, if you speak, you are hanged."

249° *Euler returns to the Russian Academy.* Euler remained at the Prussian Academy for twenty-five years, but his character did not harmonize with the type admired by Frederick the Great, and he suffered many years of petty unpleasantnesses. The Russians had held Euler in high respect, and even after he left for Prussia continued to advance him some salary. In 1760, Prussia became engaged in the Seven Years' War with Russia and other enemies. On an occasion when Frederick was away with his army at Breslau in Silesia, the Russians invaded Prussia and captured Berlin. During the course of the campaign, Euler's farm at Charlottenburg, about four miles from Berlin, was pillaged by the Russian troops. When the Russian general learned that the farm belonged to Euler, he immediately paid full indemnity, and the Russian Empress Elizabeth sent Euler an additional four thousand crowns for compensation. The warmth of the Russian feeling toward him, as contrasted with the coolness of the court of Frederick the Great, led Euler to accept an invitation from Catherine the Great in 1766 to return to the St. Petersburg Academy, where he stayed for the remaining seventeen years of his life.

250° *A narrow escape.* Five years after his return to Russia, Euler's house and its furnishings were destroyed in the great St. Petersburg fire of 1771. The now blind Euler escaped with his life only through the heroism of his Swiss servant, Peter Grimm, who, at great personal risk, carried his master through the flames to safety. Empress Catherine promptly gave Euler a new and completely furnished home.

251° *A formalist and his pencil.* A formal manipulator in mathematics often experiences the discomforting feeling that his pencil surpasses him in intelligence. This was a feeling that Euler confessed he often could not get rid of.

252° *The Euler–Diderot anecdote.* One of the most-often told anecdotes about a mathematician is the story of the discomfiture of the atheistic French philosopher Denis Diderot at the hands of the deeply religious Euler. The story was first told by Thiébault in his *Mes souvenirs de vingt ans de séjour à Berlin* of 1801, and then later retold, with significant additions, by Augustus De Morgan in his *Budget of*

Paradoxes of 1878. Since then, the story has been repeated by many authors, always in De Morgan's highly colored version. Here is the way De Morgan tells the story:

> The following story is told by Thiébault, in. his *Souvenirs de vingt ans de séjour à Berlin,* published in his old age, about 1804. This volume was fully received as trustworthy; and Marshall Mòllendorf told the Duc de Bassano in 1807 that it was the most veracious of books written by the most honest of men. Thiébault says that he has no personal knowledge of the truth of the story, but that it was believed throughout the whole of the north of Europe. Diderot paid a visit to the Russian court at the invitation of the Empress. He conversed very freely, and gave the younger members of the court circle a good deal of lively atheism. The Empress was much amused, but some of the councillors suggested that it might be desirable to check these expositions of doctrine. The Empress did not like to put a direct muzzle on her guest's tongue, so the following plot was contrived. Diderot was informed that a learned mathematician was in possession of an algebraical demonstration of the existence of God, and would give it to him before all the court, if he desired to hear it. Diderot gladly consented; though the name of the mathematician is not given, it was Euler. He advanced toward Diderot, and said gravely and in a tone of perfect conviction: *Monsieur,* $(a + b^n)/n = x,$ *donc Dieu existe; répondez!* Diderot, to whom algebra was Hebrew, was embarrassed and disconcerted, while peals of laughter rose on all sides. He asked permission to return to France at once, which was granted.

Now the De Morgan version differs from the Thiébault version in four points: (1) the mathematical formula is slightly different, (2) the mathematician is not identified by Thiébault, (3) the expression "to whom algebra was Hebrew" was added by De Morgan, (4) De Morgan has it that Diderot was mathematically unable to reply, whereas Thiébault merely says that Diderot did not reply because he sensed hostility in the audience.

The first point of difference is basically unimportant, and, as to the second point of difference, *if* the Thiébault story is true, then the mathematician *may* have been Euler, for Euler was in Russia at the time. But the third and fourth points of difference show that the De Morgan version is certainly not true. For Diderot was a very good mathematician and had published five creditable memoirs on the subject prior to his trip to Russia. For example, in his second memoir, he showed that each of the three famous problems of antiquity (the duplication of a cube, the trisection of an angle, and the quadrature of a circle) can be solved if we are given a circle and its involute. In this paper, Diderot shows a fine mastery of algebra, geometry, and calculus. It follows that the anecdote as told by De Morgan, and so often repeated by subsequent writers, is absurd.

One even wonders if the Thiébault version is true. It is curious that, if the story was believed throughout northern Europe, no one else has related it; there is, for example, no known Russian source for the story. Again, the story hardly fits Euler's character; Euler did not indulge in thoughtless and asinine behavior.

A more likely explanation is that the story originated as a canard in the court of Frederick the Great. Frederick was a bitter enemy of Diderot, and there are several sources indicating that stories about Diderot while in St. Petersburg emanated from Berlin.

One wonders how many of the "good" anecdotes about great people could completely survive the searchlight of careful historical examination.

253° *Euler's death.*   Euler retained vigor and power of mind up to the moment of his death, which occurred in his seventy-seventh year, on September 18, 1783. He had amused himself in the afternoon calculating the laws of ascent of balloons. He then dined with Lexell and his family, and outlined the calculation of the orbit of the recently discovered planet Uranus. A short time later he begged that his grandson be brought in. While playing with the youngster and sipping some tea, he suffered a stroke. His pipe fell from his hand and he uttered, "I die." At that instant, in the words of Condorcet, "Euler ceased to live and calculate."

LAGRANGE

Joseph Louis Lagrange (1736–1813) was born in Turin, Italy, into an originally French family. In 1766 he succeeded Euler at the Berlin academy, to remain there for twenty years. Later in spite of the chaotic political situation in France, he moved to Paris and assisted at the newly established École Normale and École Polytechnique. Lagrange's work had a deep influence on later mathematical research, for he was the earliest mathematician of the first rank to attempt a rigorization of the calculus. He made notable contributions in the fields of mechanics, differential equations, and the calculus of variations. He also had a penchant for number theory and wrote important papers in this field. Some of his early work in the theory of equations later led Galois to the theory of groups.

254° *Who was the most eminent mathematician of the eighteenth century?* The two most outstanding mathematicians of the eighteenth century were Euler and Lagrange, and which of the two is to be accorded first place is a matter of debate that often reflects the varying mathematical sensitivities of the debaters. Euler certainly published far more than Lagrange, and worked in many more diverse fields of mathematics than Lagrange, but he was largely a formalist or manipulator of formulas. Lagrange, on the other hand, may be considered the first true analyst and, though his collection of publications is a molehill compared with the Vesuvius of Euler's output, his work has a rare perfection, elegance, and exactness about it. Whereas Euler wrote with a profusion of detail and a free employment of intuition, Lagrange wrote concisely and with attempted rigor.

A spirited and exciting argument can be created at a sizable mathematics meeting or among the mathematics staff of a university by raising the question whether Euler or Lagrange is to be regarded as the superior mathematician. The result of the argument will, in all likelihood, exhibit an almost fifty-fifty split in the debaters, about half being supporters of Euler and half supporters of Lagrange.

Some years ago, Walter Crosby Eells endeavored to determine the one hundred most eminent mathematicians living prior to 1905, and to list these men in order of eminence. (See his paper, "One hundred eminent mathematicians," *The Mathematics Teacher*, Nov. 1962, pp. 582–588.) In this list Newton appears in first place, Leibniz in second place, Lagrange in third place, and Euler in fourth place.

Concerning Eell's list, it is only fair to quote from him about the method he employed.

Various methods, none of them entirely free from objection, have been used by different investigators to select a group of men eminent in some field, and to arrange them in order of their eminence. On the whole there is little doubt that the so-called "space method" yields the most reliable results, particularly for the selection of men no longer living and whose places therefore have become relatively fixed in history. It consists essentially in measuring the amount of space occupied by the account of the man's career in biographical diction- aries, standard encyclopedias, and other suitable reference works, and designating as "eminent" all who occupy more than a minimum space. . . .

Is the assumption that a man's eminence as a mathematician may be determined by the amount of space devoted to his life and accomplishments a satisfactory method to use? "Eminence" in itself is a difficult term to define satisfactorily and a more difficult quality to measure quantitatively. Yet there is value in making the effort, if its limitations are frankly recognized.

Other methods than the space method have been used for measur- ing degrees of eminence. Among these are number of pages of work produced or published, relative frequency in portrait catalogues, frequency of mention in selected indexes, number of works catalogued in leading libraries, comparison of favorable and unfavorable adjectives used concerning them, and a combination of qualitative judgments of a group of selected judges.

A study of the various methods which have been used by different investigators leads to the conclusion that the space method is open to the least objections. It is true that many chance or controversial factors may lead to a man receiving an amount of space out of proportion to his prominence as a mathematician. If a man has had a long and varied career, much more space may be spent in discussing it than that given to another whose work may be of more real significance in the development of mathematics. His career may be more noteworthy for variety and length than for intrinsic mathematical worth. The unfor- tunate controversy over the discovery of the calculus between Newton and Leibniz and their protagonists occupies more space in the literature than its real importance justifies, but in this case (perhaps the most striking one of this type) the additional space redounds about equally to the advantage of both men. If it were eliminated entirely, Newton's position would still be preëminent, but Leibniz would probably be placed below Lagrange and Euler, rather than above them. As a matter of fact, however, all that we are justified in concluding, consid- ering the probable errors of their positions, is that all three of these men are of practically equal rank. . . .

It is thus evident that men with eventful lives or with disputed factors involved have a better chance of occupying more space than other persons of equal or greater significance for their true mathematical contributions. Even such factors, however, add in a certain sense to their "eminence." Eminence is not necessarily synonymous with accomplishment. But on the whole, an unprejudiced reading of the histories shows that the authors of them have not been affected by such considerations sufficiently to cause serious distortion of the results summarized in the table. They have used most of their space in discussing the works of a man and his importance in the development of the science, not in considering details of his life. It seems, therefore, that by choosing a sufficiently varied group of authorities of different nationalities, as attempted in this study, a very satisfactory working list of eminent mathematicians may be selected with a considerable degree of assurance.

255° *Frederick the Great invites Lagrange to Berlin.* When Euler left Berlin, Frederick the Great wrote to Lagrange that "the greatest king in Europe" wished to have at his court "the greatest mathematician of Europe."

256° *A scientific poem.* Lagrange's monumental *Méchanique analytique*, which dates from his Berlin period, was described by Sir William Rowan Hamilton as a "scientific poem."

257° *A lofty pyramid.* Napoleon Bonaparte summed up his estimate of Lagrange by saying, "Lagrange is the lofty pyramid of the mathematical sciences."

258° *A very short paper.* Lagrange once thought that he had "proved" Euclid's parallel postulate. He wrote up his "proof," took it to the Institute, and began to read it. In the first paragraph he noted an oversight. He muttered, "I must think more on this," put the paper in his pocket, and sat down.

259° *Wealth versus mathematics.* Lagrange's father was rich, both by inheritance and by marriage. But he was an incorrigible and unwise speculator, and when the time came to bequeath his fortune to his son, very little was left. Later in life Lagrange referred to this disaster as the luckiest thing that had happened to him: "If I had inherited a fortune, I probably should not have taken up mathematics."

260° *A great waste.* Lagrange was revolted by the cruelties of the Terror that followed the French Revolution. When the great chemist Lavoisier went to the guillotine, Lagrange expressed his indignation at the stupidity of the execution: "It took the mob only a moment to remove his head; a century will not suffice to reproduce it."

261° *The marriage of September and April.* Lagrange was subject to great fits of loneliness and despondency. He was rescued from these in the twilight of his life, when he was fifty-six, by a young girl nearly forty years his junior. She was the daughter of his friend, the astronomer Lemonnier. She was so touched by Lagrange's unhappiness that she insisted on marrying him. Lagrange submitted, and the marriage turned out ideal. She proved to be a very devoted and competent companion, and succeeded in drawing her husband out and reawakening his desire to live. Of all his prizes in the world, Lagrange claimed, with honesty and simplicity, the one he most valued was his tender and devoted young wife.

LAPLACE

Pierre-Simon Laplace (1749–1827) was born of poor parents in Beaumont-en-Auge, in France. His mathematical ability early won him good teaching posts, and as a political opportunist he ingratiated himself with whichever party happened to be in power during the uncertain days of the French Revolution. His most outstanding work was done in the fields of celestial mechanics, probability, differential equations, and geodesy.

262° *Laplace seeks employment.* When Laplace arrived as a young man in Paris seeking a professorship of mathematics, he submitted his recommendations by prominent people to d'Alembert, but was not received. Returning to his lodgings, Laplace wrote d'Alembert a brilliant letter on the general principles of mechanics. This opened the door, and d'Alembert replied: "Sir, you notice that I paid little attention to your recommendations. You don't need any; you have introduced yourself better." A few days later Laplace was appointed professor of mathematics at the Military School of Paris.

263° *The Newton of France.* Laplace's great five-volume work, the *Méchanique céleste* (1799–1825), earned him the title of "the Newton of France." The work embraced all previous discoveries in this field along with Laplace's own contributions, and marked the author as the unrivaled master in the subject. The work has been called the *Almagest* of modern times.

264° *An unneeded hypothesis.* When Napoleon teasingly remarked to Laplace that God is not mentioned in the *Méchanique céleste*, Laplace replied, "Sire, I did not need that hypothesis." When Napoleon later reported this reply to Lagrange, the latter remarked, "Ah, but that is a fine hypothesis. It explains so many things."

265° *Il est aisé à voir.* Jean-Baptiste Biot (1774–1862), who helped Laplace revise his *Méchanique céleste* for the press, said that Laplace was often unable to recover the steps in his chains of reasoning, but, if satisfied that the conclusions were correct, was content to insert the often-repeated phrase, "It is easy to see."

The American astronomer Nathaniel Bowditch (1773–1838), when he translated the *Méchanique céleste* into English, remarked, "I never come across one of Laplace's 'Thus it plainly appears' without feeling that I have hours of hard work before me to fill up the chasm and find out and show *how* it plainly appears."

266° *The faith of a mechanist.* In his *Théorie analytique des probabilités* (1812), Laplace says: "Given for one instant an intelligence which could comprehend all the forces by which nature is animated and the respective positions of the beings which compose it, if moreover this intelligence were vast enough to submit these data to analysis, it would embrace in the same formula both the movements of the largest bodies in the universe and those of the lightest atom: to it nothing would be uncertain, and the future as the past would be present to its eyes. The human mind offers a feeble outline of that intelligence, in the perfection which it has given to astronomy. Its discoveries in mechanics and in geometry, joined to that of universal gravity, have enabled it to comprehend in the same analytical expressions the past and future states of the world system." Elsewhere we find, more pithily: "All the effects of nature are only mathematical consequences of a small number of immutable laws."

267° *Laplace's "stepchildren."* For all his slippery politics and his tendencies towards snobbishness, Laplace did extend sincere generosity to beginners. Biot has told that as a young man he once read a scientific paper before a session of the French Academy at which Laplace was present. Afterwards, Laplace drew him aside and showed him the identical discovery in one of his own old but still unpublished manuscripts. Cautioning young Biot to secrecy, Laplace urged him to go ahead and publish his work.

Laplace used to say that young beginners in mathematical research were his stepchildren, and there are several instances in which he withheld publication of a discovery to allow a beginner the opportunity to publish first. Sadly, such generosity is rare in mathematics.

NAPOLEON BONAPARTE

NAPOLEON Bonaparte (1769–1821) had a genuine respect for mathematics, admired the great mathematicians of his day, and was himself something of an amateur geometer.

268° *Mathematics and the welfare of the state.* Gaspard Monge (the elaborator of descriptive geometry and the father of the differential geometry of curves and surfaces in space) and Joseph Fourier (the investigator of the mathematics of heat flow and the first to show the importance of trigonometric series) were close friends of Napoleon and did much to assist the emperor in a number of his military and cultural undertakings. It has been said that Monge was perhaps the only man for whom Napoleon had a truly unselfish and abiding friendship.

Napoleon realized that ignorance leads only to annihilation. Accordingly he ordered and encouraged the creation of schools. To train and develop much-needed teachers, the École Normale was founded; to stimulate advanced study and research, the École Polytechnique was founded. Both Monge and Fourier energetically assisted Napoleon in the creation of these two great schools. Napoleon's recognition of the importance of mathematics in particular appears in his insightful statement, "The advancement and perfection of mathematics are intimately connected with the prosperity of the State."

Napoleon frequently sought enjoyment and relaxation in simple mathematical exercises. Somewhat difficult to believe, though, is the following remark made by J. S. C. Abbott in his biography of Napoleon: "When he [Napoleon] had a few moments for diversion, he not unfrequently employed them over a book of logarithms, in which he always found recreation."

269° *Napoleon's problem.* Napoleon was a friend of the Italian poet and mathematician Lorenzo Mascheroni (1750–1800). Mascheroni had shown interest in carrying out Euclidean constructions (geometric

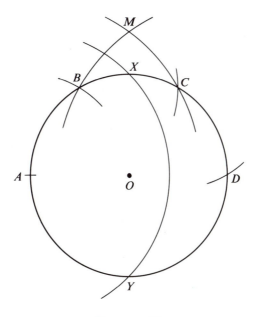

FIGURE 31

constructions that can be accomplished with straightedge and compasses) with the compasses alone, and quite likely Napoleon received his interest in such problems through his contact with Mascheroni.

Florian Cajori, in his *A History of Mathematics*, says: "Napoleon proposed to the French mathematicians the problem, to divide the circumference of a circle into four equal parts by the compasses only. Mascheroni does this by applying the radius three times to the circumference; he obtains the arcs *AB*, *BC*, *CD*; then *AD* is a diameter; the rest is obvious." The reader may care, before going on to the next paragraph, to try to complete the "obvious" part of the construction.

Though hardly obvious, the construction is not difficult. Draw an arc (see Figure 31) with center *A* and radius *AC* to intersect an arc with center *B* and radius *DB* in point *M*; draw the circle of center *A* and radius *OM* (where *O* is the center of the given circle) to cut the given circle in *X* and *Y*; then it can be shown that *A*, *X*, *D*, *Y* are vertices of a square inscribed in the given circle.

It is said that Napoleon believed there was some similarity between

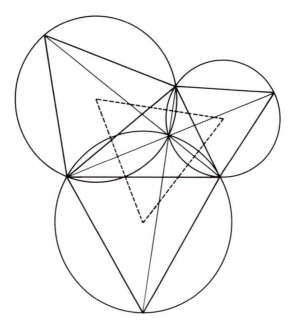

FIGURE   32

military strategy and mathematical strategy, and that on occasion he would worry his army officers by testing them with simple exercises in geometry.

270° *Napoleon's theorem.* There is a very pretty geometrical theorem that says: *If equilateral triangles are described externally on the sides of any triangle, then the centers of the circumscribed circles of these triangles form a fourth equilateral triangle* (see Figure 32). This theorem, which has many attractive generalizations and extensions, has been attributed to Napoleon, though the possibility that he knew enough geometry to discover and prove the theorem seems to be on a par with the possibility that he knew enough English to compose the famous palindrome

ABLE WAS I ERE I SAW ELBA.

QUADRANT FOUR

From some philately
to a classroom performance

ABEL AND AGNESI

271° *Honored on a postage stamp.* Niels Henrik Abel (1802–1829), by long odds Norway's greatest bid to fame in the field of mathematics, died at the very young age of twenty-six years and eight months of poverty, disappointment, malnutrition, and chest complications, while desperately seeking a modest university teaching post to hold his body and soul together. Allowed, from lack of concern and from withheld recognition, by his government to die, he finally received posthumous recognition, and Norway's greatest mathematician now appears on some small-denominational postage stamps.*

But the mathematicians, in their characteristic manner, have erected monuments to Abel that are far more lasting than even bronze. Today, anyone who reads advanced mathematical texts will encounter Abel's name perpetuated in an abundance of theorems and theories, for this is the man of whom Hermite said, "He has left mathematicians something to keep them busy for five hundred years."

272° *Crelly and Keilhau.* Abel spent his last days at Froland, in the English home of a Mr. and Mrs. Smith where his fiancée Crelly Kemp served as governess. Surrounded by solicitous attention and lovingly nursed by Crelly, assisted by the two oldest daughters of the household, he became weaker and his cough increased. His last thoughts were on the future of his beloved Crelly, and he wrote to his good

---

* Among other mathematicians appearing on postage stamps are Archimedes, Aristotle, Farkas Bolyai, János Bolyai, Boscovich, Brahe, Buffon, L. N. M. Carnot, N. L. S. Carnot, Ch'ang Hong, Ch'unh Chih, Chaplygin, Copernicus, Cristescu, Cusanus, d'Alembert, da Vinci, Descartes, de Witt, Dürer, Einstein, Euler, Galileo, Gauss, Gerbert, Hamilton, Helmholtz, Hipparchus, Huygens, Kepler, Kovalevsky, Krylov, Lagrange, Laplace, Leibniz, Liapunov, Lobachevsky, Lorentz, Mercator, Monge, Nasir-ed din, Newton, Ostrogradsky, Pascal, Poincaré, Popov, Pythagoras, Ramanujan, Riese, Stevin, Teixeira, Titeica, and Torricelli. Russia and France have been the most generous in representing mathematicians on postage stamps; England has never done so and the United States only once. Da Vinci, Galileo, Copernicus, and Einstein have each been represented by four or more different countries.

friend Baltazar Mathias Keilhau imploring him to assist Crelly when he was gone. Abel expressed—half jokingly perhaps—that the two of them should marry after his death.

Though Crelly and Keilhau had never met, a few months after Abel's death, the faithful friend, feeling a deep sense of obligation, wrote Crelly explaining that he had come to feel that the most beautiful way in which the prayer of his friend could be fulfilled would be for her to consent to become his wife. In January, 1830, upon Crelly's twenty-sixth birthday, the two became engaged. They married the next fall and Crelly became "Fru Professor" in Oslo, but not, as she had formerly and so ardently dreamed, at the side of her dear Niels Abel.

Abel was buried at Froland Church in a plot intended for the Smiths. While still engaged, Crelly and Keilhau drove to Froland Church to pay their last respects at the grave. It was on this occasion that Keilhau conceived the idea of collecting funds to erect there a lasting monument to Abel. The visitor of today pilgrimaging to Froland Church can see Keilhau's monument to his friend.

273° *A precocious and versatile somnambulist.* The most erudite woman mathematician of eighteenth-century Italy was Maria Gaetana Agnesi (pronounced än yä′ zē). She was born in Milan in 1718, the first of her father's twenty-one children. At an early age she mastered Latin, Greek, Hebrew, French, Spanish, German, and a number of other foreign languages. When she was only nine years old, her Latin discourse defending higher education for women was published. During her childhood, her father hosted gatherings of the intelligentsia at which Maria would converse with learned professors on any topics of their choice in their native languages. Later, when she was twenty, there appeared her *Propositiones philosophicae*, a series of essays on philosophy and natural science that arose from discussions at her father's gatherings. At the age of thirty she published a two-volume work on analytic geometry that enjoyed great and long popularity and was translated into several languages. The following year Pope Benedict XIV appointed her honorary faculty member of the University of Bologna, but (contrary to inaccurately told stories) she never lectured there.

Maria greatly disliked publicity and endeavored at different times to lead a secluded life. She finally succeeded when her father died in 1752, devoting the remainder of her life to charitable works and religious study. In 1771 she was appointed director of a beneficent institution in Milan, and it was there that she died in 1799. She had a younger sister, Maria Teresa Agnesi (1724–1780), who became an accomplished musician and composer.

During her lifetime, Maria Gaetana Agnesi achieved fame not only as a mathematician, linguist, and philosopher, but as a somnambulist. On several occasions she proceeded, while in a somnambulistic state, to her study, lighted a lamp, and solved some problem that she had left incomplete when awake. In the morning she would be surprised to find the solution carefully and completely worked out on paper on her desk.

274° *The witch of Agnesi.* Pierre de Fermat (1601–1663), who must be considered one of the inventors of analytic geometry, at one time interested himself in the cubic curve, which in present-day notation would be indicated by the Cartesian equation $y(x^2 + a^2) = a^3$. The curve is pictured in Figure 33. Fermat did not name the curve,

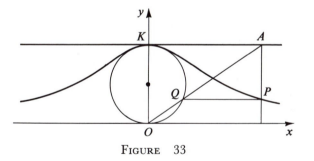

FIGURE 33

but it was later studied by Guido Grandi (1672–1742), who named it *versoria*. This is a Latin word for a rope that guides a sail. It is not clear why Grandi assigned this name to the cubic curve. There is a similar obsolete Italian word, *versorio*, which means "free to move in every direction," and the doubly-asymptotic nature of the cubic curve suggests that perhaps Grandi meant to associate this word with the curve. At any rate, when Maria Gaetana Agnesi wrote her widely read analytic geometry, she confused Grandi's *versoria* or *versorio* with *versiera*, which in Latin means "devil's grandmother" or "female goblin." Later, in 1801, when John Colson translated Agnesi's text into English, he rendered *versiera* as "witch." The curve has ever since in English been called the "witch of Agnesi," though in other languages it is generally more simply referred to as the "curve of Agnesi."

The witch of Agnesi possesses a number of pretty properties. First of all, the curve can be neatly described as the locus of a point P in the following manner. Let a variable secant OA (see Figure 33) through a given point O on a fixed circle cut the circle again in Q and cut the tangent to the circle at the diametrically opposite point K to O in A. The curve is then the locus of the point P of intersection of the lines QP and AP, parallel and perpendicular, respectively, to the aforementioned tangent. If we take the tangent through O as the x-axis and OK as the y-axis of a Cartesian coördinate system, and denote the diameter of the fixed circle by a, the equation of the witch is found to be $y(x^2 + a^2) = a^3$. The curve is symmetrical in the y-axis and is asymptotic to the x-axis in both directions. The area between the witch and its asymptote is πa^2, exactly four times the area of the fixed circle. The centroid of this area lies at the point $(0, a/4)$, one fourth the way from O to K. The volume generated by rotating the curve about its asymptote is $\pi^2 a^3/2$. Points of inflection on the curve occur where OQ makes angles of 60° with the asymptote.

An associated curve called the *pseudo-witch* is obtained by doubling the ordinates (the y-coördinates) of the witch. This curve was studied by James Gregory in 1658 and was used by Leibniz in 1674 in deriving his famous expression

$$\pi/4 = 1 - \tfrac{1}{3} + \tfrac{1}{5} - \tfrac{1}{7} + \cdots.$$

CHARLES BABBAGE

ABOUT 1812, the English mathematician Charles Babbage (1792–1871) began to consider the construction of a machine to aid in the calculation of mathematical tables. According to one story, the origin of his ideas about such machines came to him when the younger Herschel brought in for checking some calculations that had been performed for the Astronomical Society. In the course of the tedious checking, Herschel and Babbage found a number of errors, finally causing Babbage to exclaim, "I wish to God these calculations had been executed by steam." "It is quite possible," replied Herschel. From this chance interchange of remarks arose the obsession that was to rule Babbage for the rest of his life and was to transform him from a cheerful young man into a bitter old one.

With great energy and enthusiasm, Babbage threw himself into his project, investing and finally losing his own considerable personal fortune in the venture. In 1822 he pointed out, in a letter to the President of the Royal Society, the advantages to the government of a machine that would calculate the lengthy tables for navigation and astronomy, and proposed to construct such a machine for the government's use. His proposal was enthusiastically received, and in 1823 the government agreed to grant him funds for the enterprise, which was expected to take three years.

Babbage set to work to make what he called his *difference engine*, capable of employing twenty-six significant figures and of computing and printing successive differences out to the sixth order. But the work did not progress satisfactorily. Babbage was constantly dreaming up new and grander ideas about the machine that led to the scrapping of all that had already been done. The result was that after about ten years, the governmental aid was withdrawn. Babbage thereupon abandoned his difference engine and commenced work on a much more ambitious machine that he called his *analytical engine*, which was intended to execute completely automatically a whole series of arithmetical calculations assigned to it at the start by an operator. This machine, also, was never completed, largely because of lack of funds and because the necessary precision tools were as yet not made.

The British government had poured some £17,000 into the construction of the difference engine, and Babbage himself had contributed a comparable amount. The unfinished machine, and the drawings for the complete machine, were deposited in the Museum at King's College in London, and then later moved to the South Kensington Museum, where they now are. The part on exhibit is still in good working order and not long ago was taken apart, thoroughly cleaned, and reassembled so that an exact copy could be made for the museum of the International Business Machines (IBM) Corporation.

71

Though the Babbage projects failed, they did provide the inspiration for the remarkable giant mechanical and electronic brains that have come into existence in recent years. Babbage had enunciated the principles on which all modern computing machines are based. When the British magazine *Nature* published an article in 1946 discussing one of America's first large calculators (the Harvard relay computer, Mark I), it entitled the article, "Babbage's Dream Comes True."

275° *An unheralded prophet.*   Charles Babbage previsioned the vast and important field known today as *operations research*. It was his book *Economy of Manufactures and Machinery*, devoted to a study of scientific manufacturing processes of all kinds, and produced as a by-product of his interest in computing machinery, that foreshadowed the concept. Briefly put, operations research is the scientific analysis of business problems aimed at providing executives and management with information leading to more effective operation of their businesses.

Babbage was in the lead in another way. He waged a determined campaign for governmental subsidy in research and education. He did this when research was still essentially a gentleman's hobby.

276° *The Analytical Society.*   Babbage was a sociable and gregarious person who possessed a considerable sense of humor. He assisted in the founding of the Astronomical Society (1820), the British Association for the Advancement of Science (1831), and the Statistical Society of London (1834). He was brilliantly present at many dinner parties, and he belonged to a host of clubs devoted to whist, chess, boating, and other things.

Babbage's most intimate friends were the younger Herschel (later Sir John) and George Peacock (later Dean of Ely). As undergraduates, Babbage, Herschel, and Peacock entered into a compact that they would "do their best to leave the world wiser than they found it." Their first step toward achieving this goal occurred in 1812, when, together with several others, they founded the Analytical Society, whose purpose was to put "English mathematicians on an equal basis with their Continental rivals."

In short, the Analytical Society wished to remedy the severe situation into which the English mathematicians had worked themselves following the earlier bitter controversy between Newton and Leibniz over priority of the discovery of the calculus. The English mathematicians, backing Newton as their leader, had cut themselves off from continental developments, with the result that mathematics in England suffered detrimentally for almost a hundred years. While on the Continent the mathematicians were using Leibniz's much more fluent differential notation dy/dx for the derivative, the English mathematicians were clinging to Newton's far less fortunate fluxional notation $\dot{y}$ for the derivative. Accordingly, in Babbage's humorous words, the Analytical Society advocated "the principles of pure d-ism as opposed to the dot-age of the university."

In 1816, Babbage, Herschel, and Peacock translated Lacroix's elegant one-volume *Calculus* into English. This did much to aid the Society's reform in the teaching and notation of the calculus. A further gain was made in 1817, when Peacock was appointed an examiner for the mathematical tripos; he replaced fluxional symbols by differential notation on the Cambridge examinations.

277° *Babbage and Tennyson.*   Babbage was always calling attention to errors in tables and in calculations, and he frequently wrote letters to scientific societies and to government agencies pointing out such oversights. His demand for statistical accuracy even extended to poetry, and it is said that he sent the following letter to Lord Alfred Tennyson in connection with a couplet in the poet's "The Vision of Sin":

> "Every minute dies a man, / Every minute one is born": I need hardly point out to you that this calculation would tend to keep the sum total of the world's population in a state of perpetual equipoise, whereas it is a well-known fact that the said sum total is constantly on the increase. I would therefore take the liberty of suggesting that in the next edition of your excellent poem the erroneous calculation to which I refer be corrected as follows: "Every moment dies a man / And one and a sixth is born." I may add that the exact figures are 1.167, but something must, of course, be conceded to the laws of metre.

It is a fact that up to 1850 the couplet in all editions reads, "Every minute dies a man, / Every minute one is born," while in all later editions it reads, "Every moment dies a man, / Every moment one is born."

278° *The origin of miracles.*   It is well known that it is possible to define a sequence of numbers by a complicated mathematical rule such that the first hundred million terms, say, proceed according to some simple obvious pattern, the next term violently violates that pattern, and the remaining terms continue according to the pattern.* Babbage repeatedly alluded to this fact, and he described the programming of a calculator for generating such a sequence. In this argument he felt he had found a possible explanation of the origin of miracles in a world otherwise controlled by apparently simple and orderly natural laws. God emerges as a complicated master programmer.

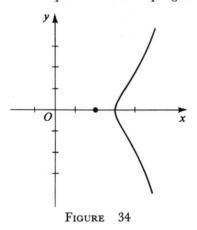

FIGURE   34

Thomas Hill has called attention to another mathematical illustration by Babbage explaining the possibility of miracles—an illustration derived from the existence of isolated points satisfying the equation of a curve. For example, the equation

$$y^2 = (x - 2)^2(x - 3)$$

---

* Consider, for example,

$$f(n) = n \prod_{i=1}^{a} (n - i) \prod_{j=a+2}^{\infty} (j - n),$$

where $a = 100{,}000{,}000$, say. Then $f(n) = n$ for all positive integral n except $n = 100{,}000{,}001$. For this value of n, $f(n) = \infty$.

74

is satisfied by all the points of the curve shown in Figure 34, and also by the isolated point (2, 0). If we should try inductively to obtain the equation of the curve, we might fail altogether to include the isolated point, which, though standing alone like a miracle to the observer of the points on the curve, is nevertheless rigorously included in the law of the curve.

279° *The scientific gadfly.* It is true that repeated disappointment turned Babbage into a bitter man and caused him to write diatribes on a number of matters. But it must be admitted, that though some of his attacks were exaggerated, his position generally was fundamentally sound and attracted a good deal of support.

One of Babbage's pet peeves concerned the neglect of science in England on the part of the Royal Society, and he submitted a plan of sweeping reform for the society. He recommended, for example, that the publication of scientific articles should be among the tests for membership, that there should be democratic procedures for elections in the society, and that there should be free debate of society policies at the meetings. The society rejected his proposals without discussion. Infuriated by this refusal, Babbage described the Council of the Royal Society as "a collection of men who elect each other to office and then dine together at the expense of the society to praise each other over wine and to give each other medals."

Babbage found a similar target in the Royal Observatory at Greenwich. He had been refused a copy of some of the Greenwich observations and then later accidentally found five tons of the tables in a shop that had bought them by the pound for making cardboard. Babbage remarked that "the Astronomer Royal is certainly the best fitted to decide what should be done with his own publications, but it is hardly possible to invent a more extravagant way of compensating a public servant than to establish an observatory and computing center for the production and printing of astronomical tables simply as a source of wastepaper." Perhaps this is not the place to comment further on the senseless waste of bureaucratic organizations.

280° *The cowcatcher and electric telegraph.* During a large dinner party, at which Babbage and several officers of the recently constructed railway were present, the conversation naturally turned to the new mode of travel, particularly upon some of its difficulties and dangers. Among other things, it was observed that obstacles might be placed, either accidentally or purposely, on the rails, and that this might lead to expensive or fatal results. To safeguard against these evils, Babbage suggested two possible remedies:

(1) That every engine should have just in advance of each of its front wheels a powerful framing, supporting a strong piece of plate-iron, descending within an inch or two of the upper face of the rail. These iron plates should be fixed at an angle of 45° with the line of the rail, and also at the same angle with respect to the horizon. Their shape should be something like that of ploughshares, and their effect would be to pitch any obstacle obliquely off the rail.

(2) Place in front of each engine a strong leather apron attached to a powerful iron bar, projecting five or six feet in front of the engine and about a foot above the ballast. The effect of this would be that any animal straying over the railway would be pitched into this apron, probably having its legs broken, but forming no impediment to the progress of the train. If used on enclosed roads, it still might occasionally save the lives of incautious persons, although possibly at the expense of broken limbs.

One day Babbage found himself seated at dinner next to an eminent London banker. The new system of railroads entered as a topic of conversation and much was said in its favor. But the banker did not appear to agree. At length Babbage asked for the banker's opinion. "Ah," said the banker, "I don't approve of this new mode of traveling. It will enable our clerks to plunder us, and then be off to Liverpool on the way to America at the rate of *twenty* miles an hour." Babbage suggested that science might perhaps remedy this evil, and that "possibly we might send lightning to outstrip the culprit's arrival at Liverpool, and thus render the railroad a sure means of arresting the thief." At the time Babbage uttered these words, he had no idea of how soon they would be realized.

281° *Legibility of mathematical tables.* Babbage's *Table of Logarithms* was published in London in 1826. In the preparation of the first edition of these tables, Babbage and the printer carried out one of the first extensive and systematic investigations into the various factors that make printed tables both easily legible and optically enduring. In the preface to the work Babbage summarizes the studies and experiments with sizes and arrangement of type and with the color and texture of paper.

Later, in 1831, in a further effort to determine the easiest to read, Babbage had printed a single copy of some pages of his table of logarithms in 21 volumes on 151 variously colored papers with ten different colors of ink (light blue, dark blue, light green, dark green, olive, yellow, light red, dark red, purple, and black), and also in metallic gold, silver, and copper on vellum and different thicknesses of paper.

It may be that today one cannot achieve the fine understanding between author and printer that existed in the days of hand-set type; something in the art of printing has become lost and overruled by matters of ease of production.

282° *Organ grinders.* Arthur Schopenhauer, the bitter and misanthropic German philosopher, developed an inordinate hatred of cabmen and the sharp cracking of their whips. In the same way, Babbage developed an excessive and unrestricted hatred for organ grinders and embarked upon a lifelong battle against them. He sued them, hailed them into court, and publicly abused them on every possible occasion. His one-man fight against street musicians brought him, at least in London, as much fame as all his scientific accomplishments combined. He claimed that his creative ideas vanished as soon as an organ grinder began to play, and he calculated that such interruptions destroyed one fourth of his working power. A magistrate once asked him if he seriously believed that a man's brain would be impaired by listening to a hand organ, and Babbage replied, "Certainly not, for the obvious reason that no man having a brain ever listened to street musicians." It has been said that Babbage was the mathematical Timon of his age, for he hated mankind in general, Englishmen in particular, and organ grinders most of all.

The lightness with which some took Babbage's battle is clear from his obituary notice in the London *Times* in 1871, which remarked that Babbage lived to be almost eighty, "in spite of organ-grinding persecutions."

SOME B'S

283° *John Taine's review of a book by Eric Temple Bell.* One Sunday in 1951 the *Pasadena Star-News* carried, in its book review column, an enthusiastic review of a new book, *Mathematics, Queen and Servant of Science*, by Eric Temple Bell (1883–1960), at that time Caltech's renowned mathematician. The review was written by John Taine and in part read: "The last flap of the jacket says Bell 'is perhaps mathematics' greatest interpreter.' Knowing the author well, the reviewer agrees."

John Taine should have known the author well, for he *was* Eric Temple Bell. "John Taine" was the pen-name Bell assumed when he wrote his highly successful science fiction novels of the 1920's and 30's.*

284° *In memory of three apples.* Farkas (in Latin, Wolfgangus, and in German, Wolfgang) Bolyai (1775–1856) spent the three years from 1796 to 1799 as a student at Göttingen University, becoming a very intimate friend of his fellow student Carl Friedrich Gauss. The two spent much time in each other's company, exchanged mathematical ideas, and took long rambles together. Farkas Bolyai is particularly remembered today as the father of János Bolyai, one of the earliest independent discoverers of a non-Euclidean geometry. He himself became a professor of mathematics, physics, and chemistry in the college at Maros-Vásárhely, and he wrote a number of works on elementary mathematics. Indeed, it was in an appendix to one of Farkas Bolyai's works that his son's ideas on non-Euclidean geometry were published, and there is no doubt that it was he who first interested his son in Euclid's parallel postulate.

Two remarks made by Farkas Bolyai have often been repeated. The first is his fine tribute to Gauss, whom he called "the mathematical giant who, from his lofty heights, embraces in one view the stars and the abysses." The second is his modest request that no monument mark his grave, only an apple tree in memory of the three

---

* See, for example, *The Time Stream*, *The Greatest Adventure*, *The Purple Sapphire*, *Seeds of Life*, and *White Lily*, all reprinted by Dover Publications, Inc.

apples: the two of Eve and Eris, which made the earth an inferno, and that of Newton, which elevated the earth again into the family of the heavenly bodies.

285° *Mathematician, violinist, fencer.* János Bolyai (1802–1860), famous son of Farkas Bolyai and one of the discoverers of non-Euclidean geometry, was educated for a military career, and distinguished himself as a creative mathematician, an impassioned violin player, and an expert swordsman. It is told that when in garrison with fellow cavalry officers, he accepted a challenge to duel thirteen of the officers on the condition that after each duel he be permitted to play a piece on his violin. According to the story, he came out victor in all thirteen duels.

The American mathematician George Bruce Halsted (1853–1922) declared János Bolyai's *Science of Absolute Space* to be "the most extraordinary two dozen pages in the history of thought." This tract, which constituted János Bolyai's published treatment of a non-Euclidean geometry, appeared in 1832 in the form of an appendix to a semi-philosophical work on elementary mathematics written by his father. Though János Bolyai's ideas were not published until 1832, as early as 1823 he began to understand the true nature of the new geometry, and in a letter to his father he exclaimed, "Out of nothing I have created a strange new universe!"

286° *Euclid as a physician.* Bernhard Bolzano (1781–1848), who in his day was a frowned-upon Czechoslovakian priest and an overlooked mathematician, has told a little anecdote about himself in which Euclid played the role of a physician. Bolzano was on vacation in Prague when he was attacked by an illness that manifested itself in bodily chills and painful weariness. To take his mind from his condition, he picked up Euclid's *Elements* and for the first time read the masterly exposition of the Eudoxian doctrine of ratio and proportion set out in Book V. The ingenuity of the treatment filled him with such vivid pleasure that, he said, forthwith he felt as well as ever.

287° *A precocious arithmetician.* Nathaniel Bowditch (1773–1838) was a self-made mathematician who achieved an international reputation through his highly useful *New American Practical Navigator*

(1802) and his very meritorious translation of Laplace's *Mécanique céleste* (published in 1829–1839). It was at the age of seventeen that he began a serious study of Latin so that he might read Newton's *Principia*, and he later mastered French, Italian, and German in order to study mathematics in those languages also.

Nathaniel's precocity in arithmetic is evidenced by the following anecdote. As a small boy Nathaniel attended a school run by a Master Watson. One day Master Watson put a long arithmetic problem on a big boy's slate. Nathaniel begged for a similar problem but was informed he was too small. That evening Nathaniel complained to his parents, who wrote Master Watson a note in their son's defense. So the next day the irritated Watson put a *very* long problem on Nathaniel's slate, saying, "There, that will keep you busy for a while." Nathaniel quickly solved and double checked the problem and asked for another.

"Too hard for you, eh?" taunted Watson. "No, you may not have another problem. Work on the one you have until you get the answer."

"But I have the answer," replied the boy.

"What? Who helped you?"

"No one."

This caused Master Watson to slap his hand hard on the desk and say, "Don't lie to me! Who helped you?"

Nathaniel steadfastly denied he had received any help, and so a beating was promised on the morrow if the boy did not tell the truth by then. That evening the worried youngster recounted the event to his family, and on the morrow his older brother, Habakkuk, accompanied him to school to explain to Master Watson that Nathaniel had really solved the problem by himself.

"He never worked that problem that fast without help! And no one can make me believe he did," expostulated Master Watson.

"Why don't you give him another, and stand over him while he works it?" suggested Habakkuk.

The suggestion was followed, and Master Watson received the surprise of his life as he observed Nathaniel very quickly solve and double check the assigned problem.

"Nathaniel," he said, "if you only knew half as much Latin as you know arithmetic, you could enter Harvard tomorrow."

288° *With an interest in ships.* Nathaniel Bowditch was not only a precocious arithmetician, but from early life on was concerned with ships and sailing. In school one day, Master Watson asked young Nathaniel, "What happened on April 19, 1775?" This, of course, was the date of the important Battle of Lexington.

"On April 19, 1775," replied Nathaniel, in all innocence, "my father's sloop, *Polly*, went aground on Anguilla Reef."

289° *A warning.* It sometimes seems, these days, that the college professors are far more conscientious than their students. Students will cut classes on the slightest pretext, whereas a professor will bear with all sorts of physical and other distresses rather than miss this appointment with his class. The dangers of this conscientiousness are brought out in the life of George Boole (1815–1864), whose career was suddenly cut short in the midst of creative labors. One day in 1864, in a drenching rain, he walked the two miles from his residence to the college (Queen's College at Cork, Ireland) where he taught, and lectured in his soaked clothes. As a result, a feverish cold fell upon his lungs and terminated his life on December 8, 1864, in the fiftieth year of his age.

In 1901, Bertrand Russell wrote, "Pure mathematics was discovered by Boole in a work which he called *The Laws of Thought....* His work was concerned with formal logic, and this is the same thing as mathematics." It was George Boole's early work in symbolic logic that blossomed later, in 1910, into the great *Principia mathematica* of Whitehead and Russell.

290° *Pi by probability.* The French naturalist Comte de Buffon (1707–1788) is known to mathematicians for two contributions—a translation into French of Newton's *Method of Fluxions*, and his "Essai d'arithmétique morale," which appeared in 1777 in the fourth volume of a supplement to his celebrated multi-volume *Histoire naturelle*.

It is in his "Essai" that Buffon suggested what was then the essentially new field of geometrical probability. As an example, Buffon devised his famous *needle problem* by which π may be determined by probability methods. Suppose a number of parallel lines (see Figure 35), distance *d* apart, are ruled on a horizontal plane, and suppose a

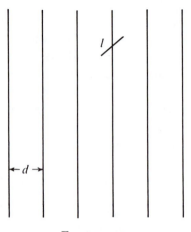

FIGURE   35

homogeneous rod of length $l < d$ is dropped at random onto the plane Buffon showed that the probability* that the rod will fall across one of the lines in the plane is given by

$$p = 2l/\pi d.$$

By actually performing this experiment a given large number of times and noting the number of successful cases, thus obtaining an empirical value for p, we may use the above formula to compute π. The best result obtained in this way was given by the Italian, Lazzerini, in 1901. From only 3408 tosses of the rod he found π correct to six decimal places! His result is so much better than those obtained by other experimenters that it is sometimes regarded with suspicion.

There are other probability methods for computing π. Thus, in 1904, R. Chartres reported an application of the known fact that if two positive integers are written down at random, the probability that they will be relatively prime is $6/\pi^2$.

---

* If a given event can happen in h ways and fail to happen in f ways, and if each of the $h + f$ ways is equally likely to occur, the *mathematical probability* p of the event happening is $p = h/(h + f)$.

CARLYLE AND LEGENDRE

291° *Thomas Carlyle's geometrical solution of quadratic equations.*
Thomas Carlyle (1795–1881), the eminent Scottish litterateur, early
in life tried to make a living as a teacher of mathematics. He came to
loathe this work, once bitterly remarking: "Teaching school is but
another word for sure and not very slow destruction." Nevertheless,
before withdrawing from teaching he left two imprints upon the history
of elementary mathematics. One of these, which will be considered in
the following Item, was a meritorious and subsequently very influential
translation of Legendre's *Éléments de géométrie* into English. His other,
and more minor, contribution was the discovery of a singularly neat
geometrical solution of quadratic equations. This appeared in one of
the editions of John Leslie's *Elements of Geometry* with the remark: "The
solution of this problem, now inserted in the text, was suggested to me
by Mr. Thomas Carlyle, an ingenious young mathematician, and
formerly my pupil."

Slightly modernized in explanation Carlyle's geometrical solution
of quadratic equations is as follows: Let us be given a quadratic equation $x^2 - gx + h = 0$. On a rectangular Cartesian frame of reference
plot the points B: $(0, 1)$ and Q: (g, h). Draw the circle on BQ as diameter and let it cut the x-axis in M and N. Then the signed lengths
of OM and ON represent the (real) roots of the given quadratic equation. The reader may care to find a proof of the correctness of this
construction. Figure 36 shows the method applied to the particular
numerical quadratic equation $x^2 + x - 6 = 0$ (for which $g = -1$,
$h = -6$). As other simple applications, the reader can consider
$x^2 - 7x + 12 = 0$ and $x^2 + 4x - 21 = 0$.

It was Professor Leslie of the University of Edinburgh who
discovered Carlyle's mathematical talent, and he did all he could to
help his young student. Carlyle never ceased to speak of Professor
Leslie with gratitude. He once wrote that his progress in mathematics
was "due mainly to the accident that Leslie alone of my professors had
some genius in his business and awoke a certain enthusiasm in me. For
several years geometry shone before me as the noblest of all sciences, and
I prosecuted it in all my best hours and moods." Carlyle's success as a

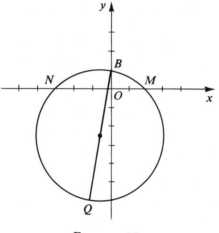

FIGURE   36

student of mathematics is attested by the fact that he took the "Dux" Prize in the mathematical class.

In the library at Brown University, in Providence, Rhode Island, is a copy of Thomas Simpson's *The Doctrine and Application of Fluxions* (1776) bearing at the top of the title page the neatly written signature: "Thomas Carlyle, Studt. Edinb. 1814."

292° *Legendre, Thomas Carlyle, and America.* Adrien-Marie Legendre (1752–1833) was born in Toulouse, France. He is known in the history of elementary mathematics principally for his very popular and influential *Éléments de géométrie,* in which he attempted a pedagogical improvement of Euclid's *Elements* by considerably rearranging and simplifying many of the propositions. His chief work in higher mathematics centered about number theory, elliptic functions, and the method of least squares. He was also an assiduous computer of mathematical tables.

Legendre's *Éléments* won high regard in continental Europe and was so favorably received in the United States that it became the prototype of the elementary geometry textbooks in this country. The first English translation was made in the United States in 1819 by John Farrar of Harvard University. The next English translation was made in 1822 by the famous Scottish litterateur, Thomas Carlyle, who early in life was a teacher of mathematics. Carlyle's translation, as later revised by Charles Davies, and later still by J. H. Van Amringe, ran through thirty-three American editions. Still other translations followed that of Carlyle.

Probably very few teachers, and perhaps no students, of high school geometry of the past few generations in the United States have realized why our geometry textbooks have differed so much, in the order and the proof of the propositions, from the arrangement and development in Euclid's *Elements*. Legendre and Carlyle greatly affected the teaching of geometry in the United States. This influence has only very recently, through the textual materials of present-day writing groups, experienced any sensible degree of remodeling.

MATHEMATICIANS AND NATURE LOVERS

LET me, for a moment, step out of my role as mere compiler of these mathematical stories and anecdotes and become a bit more personal. Having associated from early years with two particular classes of scholars—botanists (or, more widely, nature lovers) and mathematicians—I came to notice, and through the years have confirmed, a striking general difference between these two classes. The botanists are usually the most pleasant sort of people to be with; they radiate gentle modesty, are open-minded, enjoy one another's company, are kind in their professional comments about one another, and are found interesting by their nonbotanical friends. The mathematicians, on the other hand, are too often unpleasant to be with; they frequently exude self-importance, are professionally opinionated, tend to bicker and quarrel among themselves and to say unkind things of one another, take an almost gleeful pleasure in unearthing an error in another's work, and are often quite boring to their nonmathematical acquaintances.

The unpleasant features of the mathematical group are noticeable even among some of the more gifted high school students of the subject, become sharper among the college graduate students of mathematics, and often attain an undignified aspect among college instructors and professors of mathematics. I dare say the history of scholarship reveals more bitter and senseless quarrels within the mathematical fraternity than within any other scholarly group.

Thirty-some years ago, when I was a graduate student of mathematics at Harvard University, I had a dear and sage-like friend in the form of a fellow student with whom I often conversed and rambled. In one of our walks, I said to my friend: "Tell me, why are botanists so much nicer to get along with than mathematicians?" I cannot recall receiving an answer at the time, but a couple of summers ago we two met again after three decades of separation. To my surprise, my friend reminded me of my question of so many years ago, and I asked him if he had arrived at an answer. Part of his answer was this:

"Inherent in the study of mathematics today there is an element which makes for conceit. Do not misunderstand me, I do not mean that all mathematicians are conceited. I mean that the study of mathematics, especially in the higher echelons—stressing as it must a genius for analysis and abstraction, intellectual self-sufficiency, the possibility of perfect achievement, and an almost Jehova-like power to create concepts and systems—together, perhaps, with the extraordinary power that the subject seems to have in shaping our technological civilization —somehow tends to encourage the element of latent, unconscious self-conceit we are all born with."

As for the botanists, it seems quite reasonable that a close association with the intricacies and mysteries of nature would perhaps tend to engender a spirit of modesty, awe, and self-effacement.

293° *Driven to the insane asylum.*   A pathetic illustration of unpleasantness among mathematicians is the quarrel that took place between Leopold Kronecker and Georg Cantor concerning the *actual* infinite as opposed to the merely *potential* infinite. Georg Cantor (1845–1918) was the originator of set theory and the study of transfinite numbers. Kronecker, as a confirmed finitist, endeavored to show the mathematical world that Cantor's work, though it might be regarded as mysticism or as theology, certainly was not mathematics.

The two combatants were quite unequal in fighting ability, Kronecker being highly competent in verbal warfare and Cantor singularly weak. Though Kronecker's bitter animosity toward Cantor had a scientific basis, it all too often degenerated into vituperative personal invective. There were times during the onslaught when the timid, hypersensitive Cantor began to doubt himself, and in the spring of 1884, in his fortieth year, he experienced the first of the series of breakdowns that were to recur throughout the rest of his long life and to make him an intermittent inmate of the mental hospital at Halle. Each time upon emerging from the hospital, Cantor found his mental faculties extra sharp, but then another decline would set in. Cantor died in the mental hospital in Halle in 1918, at the age of seventy-three. The bulk of his professional life had been spent teaching in an inferior institution at Halle, for his dream of becoming a faculty member at the University of Berlin was constantly shattered by the fury of his foe Kronecker.

Out of the Kronecker–Cantor battle evolved the twentieth-century controversy between the intuitionist and formalist schools of philosophy of mathematics. In defense of Cantor, David Hilbert (the leader of the formalist school) proclaimed: "No one shall expell us from the paradise which Cantor has created for us." And down through time will ring Cantor's famous aphorism: "The essence of mathematics lies in its freedom."

294° *Mathematics and theology.* There may be something in Kronecker's charge that Cantor's work was more theology than mathematics. It is a fact that early in life Cantor developed a deep interest in medieval theology, and had he not become a mathematician he would in all likelihood have left his mark on philosophy or theology.

Hilbert, founder of the formalist school of philosophy of mathematics, hoped to save all classical mathematics from the ravages and attacks of the intuitionist school by establishing its inner consistency. Much as one may prove, by the rules of a game, that certain situations cannot occur within the game, Hilbert hoped to prove, by a suitable set of rules of procedure, that contradictory results can never occur within classical mathematics. For certain small parts of mathematics, Hilbert was able to carry out such consistency proofs, thus illustrating what he would like to have done for all classical mathematics, but, for the system *in toto*, unforeseen difficulties arose and the problem of consistency remained refractory.

As a matter of fact, the Hilbert program (at least in the form originally envisioned by Hilbert) appears to be doomed to failure. This truth was brought out in 1931 by the Austrian mathematician and logician Kurt Gödel, who showed, by methods acceptable to both the formalist and intuitionist schools, that it is impossible for a sufficiently rich formalized deductive system, such as Hilbert's system for all classical mathematics, to prove consistency of the system by methods belonging to the system. This remarkable result is a consequence of an even more fundamental one; Gödel proved the incompleteness of Hilbert's system—that is, he established the existence within the system of "undecidable" problems, of which consistency of the system is one. The whole matter was neatly put by F. De Sua in 1956 as follows:

> Suppose we loosely define a *religion* as any discipline whose foundations rest on an element of faith, irrespective of any element of reason which may be present. Quantum mechanics for example would be a religion under this definition. But mathematics would hold the unique position of being the only branch of theology possessing a rigorous demonstration of the fact that it should be so classified.

295° *The neglect and loss of manuscripts.* A sad, unfair, and frustrating result of the self-importance of some great mathematicians is the neglect, and sometimes loss, of certain significant manuscripts submitted to them for review. Epoch-making papers by Niels Abel and Évariste Galois, for example, met such a fate. These papers had been sent to the French Academy for judgment, and the authors waited and waited in vain to hear reports on their work. The papers were pigeonholed, and finally lost, by Cauchy, the selected referee of the Academy.

Augustin-Louis Cauchy (1789–1857) was a highly creative mathematician; he wrote extensively and profoundly in both pure and applied mathematics, and it might be that he ranks next to Euler in volume of output. His numerous contributions to advanced mathematics include researches in convergence and divergence of infinite series, real and complex function theory, differential equations, determinants, probability, and mathematical physics.

Cauchy was an ardent partisan of the Bourbons and, after the revolution of 1830, was forced to give up his professorship at the École Polytechnique and was excluded from public employment for eighteen years. Part of this time he spent in exile in Turin and Prague, and partly in teaching in some church schools in Paris. In 1848 he was allowed to return to a professorship at the École Polytechnique without having to take the oath of allegiance to the new government.

Throughout his life Cauchy was an indefatigable worker and it is regrettable that he possessed a narrow conceit and often ignored the meritorious efforts of younger and promising men.

296° *The well-balanced mathematician.* Of the great English mathematician Arthur Cayley (1821–1895) a number of interesting remarks can be made.

1. Cayley's mathematical productivity was so great that either he or Cauchy ranks next to Euler in volume of output. It might take an actual counting of pages of publication to determine which of these two men holds the singular honor.

2. Cayley did not start off making his living as a mathematician, but spent fourteen years as a lawyer before accepting the Sadlerian professorship at Cambridge. But he was always careful to restrict his law work so that it would not interfere with his mathematical interests. During his fourteen years in the practice of law, Cayley published between two and three hundred papers in mathematics.

3. Cayley developed an unusual avidity for novel reading. He read novels while traveling, while waiting for meetings to start, and at any odd moments that presented themselves. During his life he read thousands of novels, not only in English, but also in Greek, French, German, and Italian.

4. Cayley was, in the true British tradition, an amateur mountain climber, and he made frequent trips to the Continent for long walks and mountain scaling. A story has been told that he claimed the reason he enjoyed mountain climbing was that, although the ascent was arduous and tiring, the feeling of exhilaration attained when he conquered a peak was exactly like that he experienced when he solved a difficult mathematical problem or completed an intricate mathematical theory. Moreover, he said, it was easier to attain the desired feeling by climbing a mountain.

5. In 1842, at the age of twenty-one, Cayley graduated from Trinity College, Cambridge, as senior wrangler in the mathematical tripos,

and in the same year he was placed first in the even more difficult test for the Smith's prize.

6. Cayley took a deep delight in painting, especially water colors, and he exhibited a marked talent as a water-colorist.

7. In his presidential address of 1883 to the British Association for the Advancement of Science, Cayley expressed the following remarks: "It is difficult to give an idea of the vast extent of modern mathematics. The word 'extent' is not the right one: I mean extent crowded with beautiful detail—not an extent of mere uniformity such as an objectless plain, but of a tract of beautiful country seen at first in the distance, but which will bear to be rambled through and studied in every detail of hillside and valley, stream, rock, wood, and flower. But, as for everything else, so for a mathematical theory—beauty can be perceived but not explained." These remarks are not just academically-composed sentences, but reflect an actual close acquaintance with nature.

8. Cayley was the most even-tempered and least vitriolic of mathematicians. But, then, he was not only a mathematician; he was a combination of mathematician and nature lover.

297° *An appreciator rather than a creator.* August Leopold Crelle (1780–1856) was an appreciator of mathematics rather than a creator of mathematics. By profession he was a civil engineer and an architect; he built the first railroad in Germany and made a comfortable fortune. In his leisure he pursued mathematics for pleasure. Although he wrote a number of mathematical papers, his greatest contribution to the advancement of mathematics was his founding of the *Journal für die reine und angewandte Mathematik* (Journal for Pure and Applied Mathematics), more popularly referred to as *Crelle's Journal*. This was the first periodical in the world devoted exclusively to mathematical *research*, and it has appeared regularly four times every year since 1826 to the present. It tottered and almost fell in the European chaos following World War I, but was valiantly sustained by subscribers unwilling to see the great monument collapse.

Crelle had an unerring instinct in picking collaborators and recognizing true ability in mathematics in others. Here was a man who proved to be kindness itself to poor struggling Abel—and other young mathematicians. Crelle met Abel when the latter arrived in Berlin in 1825, and they had a long talk together about mathematics. In their conversation, Abel pointed out to Crelle some oversights in one of Crelle's recent articles. Crelle showed his greatness by trying to profit by Abel's remarks rather than showing indignation that so young a man should have the presumptuousness to try to pick holes in his work.

Crelle recognized immediately that Abel was far beyond him in mathematics, and he endeavored to do all in his power to assist the young man. He introduced Abel everywhere and published many of Abel's papers in his journal. It was through Crelle's strenuous efforts that finally a position on the faculty of the University of Berlin was found for Abel, but the good news tragically reached Norway two days after Abel was dead.

Crelle, by his kindness to young mathematicians, did more good for the progress of nineteenth-century mathematics than half a dozen learned academies. But Crelle was not really a mathematician; he was an engineer and an architect with only a self-effacing amateurish interest in mathematics.

CLIFFORD AND DODGSON

298° *Strong man and entertainer of children.* William Kingdon Clifford (1845–1879) was one of England's most promising mathematicians of the nineteenth century, but his death from pulmonary disease at the early age of thirty-four interrupted the complete maturing of his genius. He possessed unsurpassed powers as a teacher, was a singularly lucid expositor and an admirable orator, and mastered foreign languages with ease. It was said that during his lifetime he was the only mathematician in England who could discourse on mathematics to a general audience and make his listeners feel they understood the subject.

Clifford, like his contemporary Lewis Carroll (Charles Lutwidge Dodgson, mathematical lecturer at Christ Church, Oxford), delighted in amusing children. Teaching children to fly kites and conducting a children's party were among his greatest pleasures. To entertain the young folks he wrote a collection of fairy tales entitled *The Little People*. He also wrote many poems, only a few of which have been published.

When he presented the novelist George Eliot with a copy of *The Little People*, he enclosed the following verses:

> Baby drew a little house,
> Drew it all askew;
> Mother saw the crooked door
> And the window too.

> Mother heart, whose wide embrace
> Holds the hearts of men,
> Grows with all our growing hopes,
> Gives them birth again.

> Listen to this baby-talk:
> 'Tisn't wise or clear;
> But what baby-sense it has
> Is for you to hear.

Perhaps the most remarkable thing about Clifford was his extraordinary strength. He was able, for example, to chin the bar single-handed with either hand. He also seemed to be absolutely nerveless and did frightening things at great heights, as when he hung by his toes on the cross bars of the weathercock on a high church tower. His great strength, coupled with an equally great energy, caused him to overtax his frail health and hastened his early end.

299° *Gentle man and entertainer of children.* Charles Lutwidge Dodgson (1832–1898), author of the famous *Alice in Wonderland* and *Alice Through the Looking Glass*, and better known throughout the world by his pseudonym, Lewis Carroll, was an English mathematician and logician holding a mathematical lectureship at Christ Church, Oxford. Many people, literarily acquainted with Lewis Carroll, do not know of Dodgson the mathematician and of the fact that he published a number of texts in the field of mathematics. There is a story that Queen Victoria was so struck with Lewis Carroll's Alice books that she sent out a courtier to bring back a copy of every other book that man had written, and the courtier returned with a bundle of mathematics books that the poor Queen could not read.

Dodgson carried the art of nonsense-writing to a peak, and there are numerous instances in his literary works of remarkably involved syllogisms of logic. There is some evidence that the changes in Alice's size and proportions in the wonderland adventure form a closed set of projective transformations, and there is no doubt that *Through the Looking Glass* is based upon an end-game of chess.

How different in personality Dodgson was from Clifford! Clifford was very outgoing and was a fluent speaker; Dodgson was shy and was afflicted with a stammer. It was perhaps at least partly because of his stammer that Dodgson was drawn to the society of children, especially little girls, in whose company he felt at ease. He became an outstanding photographer of young children.

But, like Clifford, Dodgson, too, enjoyed children's parties. There once was a children's party held in a house in London, and next door there happened to be at the same time an adults' party. To amuse the children, Dodgson decided on his arrival to walk in on all fours. Unfortunately he crawled into the parlor of the wrong house.

Dodgson's gentleness was reflected in his humanity towards animals. He wrote the following consoling note to a friend of his whose pet dog had died:

> I am very sorry to hear of your sad loss. Well, you have certainly given to *one* of God's creatures a *very* happy life through a good many years—a pleasant thing to remember.

Teachers of mathematics will recall the Mock Turtle of *Alice in Wonderland*, whose "regular course" in school contained, among other subjects,* "the different branches of Arithmetic—Ambition, Distraction, Uglification, and Derision." And all problemists have, in their collections, Lewis Carroll's *Pillow Problems* and *A Tangled Tale.*†

CALCULATING PRODIGIES

Occasionally there have appeared lads possessing surprising power at mental calculation. Given large numbers these youths would very

---

* Reeling, Writhing, Mystery (ancient and modern), Seaography, Drawling, Stretching, and Fainting in Coils.

† Both books, bound as one, have been reprinted by Dover Publications, Inc.

quickly multiply them together, raise them to powers, extract their roots, and sometimes find their factors. Some of them would quickly answer more involved questions dealing with compound interest, annuities, and calendars. In most cases the lads were illiterate and employed rules of calculation of their own invention. Of course there have been several professional mathematicians who at one time or another during their lives showed remarkable ability at mental calculation. One thinks, for example, of Gauss and Euler. André Marie Ampère, the mathematical-physicist, learned when a child to perform long mental calculations. John Wallis, a predecessor of Newton and Savilian professor at Oxford, developed for his own amusement a remarkable ability of this sort in *middle age*. Here we shall confine ourselves to just a few outstanding self-taught calculators, who developed their amazing abilities in youth, and who did not later become professional mathematicians.

300° *Son of a small farmer.*   Very remarkable was Zerah Colburn, who was born in Vermont in 1804 as the son of a small farmer, and who died at the early age of thirty-six in 1840. When less than six years old he displayed his extraordinary calculating ability in a tour in America. Two years later he was taken to England where he gave many performances. He was able to give instantly the product of any two 4-digit numbers. Asked to raise 8 to the 16th power, in a few seconds he gave the correct answer, 281,474,976,710,656. He raised the numbers 2, 3, · · · , 9, to the 10th power so fast that the judge taking down his answers couldn't keep up with him. He was able instantly to give the square and cube roots (when they were integers) of large numbers—like the square root of 106,929 and the cube root of 268,336,125. Most remarkable was his ability to find factors of numbers. Asked for those of 247,483 he replied 941 and 263; for those of 171,395 he gave 5, 7, 59, and 83; asked for those of 36,083 he replied that there aren't any. He seemed equal to buffoons in his audience; asked how many black beans are required to make three white ones, he at once replied, "Three, if you skin them." One wonders if this was a prearranged part of the show. Before he died he wrote his autobiography, containing an account of some of the methods he employed.

301° *Son of a stone mason.* One of the most interesting of the calculating prodigies was George Parker Bidder, born in 1806 in Devonshire, England, where his father was a stone mason. He died in 1878. He retained his powers throughout his life and succeeded better than any other calculating prodigy in giving an account of his methods.

By the time he was nine, George's local reputation reached the point where his father found it profitable to take him about the country to exhibit his powers. By the age of thirteen he was able to give immediate answers to problems in compound interest and annuities. On one occasion he was pitted against Colburn, and generally proved to be the abler calculator. He was not, though, as good as Colburn in finding factors of numbers.

Here are some of the questions asked of Bidder at his performances. When nine he was asked: If the moon is 123,256 miles from the earth, and sound travels at the rate of 4 miles a minute, how long would it be before inhabitants of the moon could hear the battle of Waterloo? Answer: 21 days, 9 hours, 34 minutes, given in less than one minute. When ten, he was asked: How many hogsheads of cider can be made from a million apples, if 30 apples make one quart? Answer: 132 hogsheads, 17 gallons, 1 quart, and 10 apples left over, given in 35 seconds. If a coachwheel is 5 feet and 10 inches in circumference, how many times will it revolve in running 800,000,000 miles? Answer: 724,114,285,704 times with 20 inches remaining, given in 50 seconds. What is the square root of 119,550,669,121? Answer: 345,761, given in 30 seconds. When he was eleven, the astronomer Sir William Herschel asked him: Assuming light travels from the sun to the earth in 8 minutes, and that the sun is 98,000,000 miles away, if light takes 6 years and 4 months traveling from the nearest fixed star to the earth, what is the distance of that star, reckoning a year as 365 days and 6 hours and a month as 28 days? Answer: 40,633,740,000,000 miles. When twelve, he was asked: If the pendulum of a clock vibrates $9\frac{3}{4}$ inches in a second, how many inches will it vibrate in 7 years, 14 days, 2 hours, 1 minute, and 56 seconds, each year containing 365 days, 5 hours, 48 minutes, 55 seconds? Answer: $2,165,625,744\frac{3}{4}$ inches, given in less than one minute. When thirteen, he was asked: Find a number whose cube less

95

19 multiplied by its cube shall be equal to the cube of 6. Answer: 3, given instantly. If you are now 14 years old and live 50 years longer and spend half-a-crown a day, how many farthings will you spend in your life? Answer: 2,805,120, given in 15 seconds.

As with Colburn, there were attempts to be witty at Bidder's expense, but, like Colburn, he seemed able to hold his own. Asked at one of his performances in 1818 how many bulls' tails are needed to reach the moon, he immediately replied, "One, if it is long enough."

There were other members of the Bidder family possessing similar powers. One of George's older brothers became an actuary, and when his books were destroyed by fire he rewrote them in six months from memory, dying, it is said, of consequent brain fever. Another older brother memorized the entire Bible and was able to give chapter and verse of any quoted text. George's first son, who became a well known lawyer, was able to find the product of two fifteen-digit numbers, but in neither accuracy nor speed was he equal to his father. All these Bidders possessed phenomenal memories. As an example of George's remarkable memory we have the instance when, in 1816, a number was read to him backwards and he at once gave it in normal form. An hour later, when asked if he remembered the number, he immediately repeated it. The number was

$$2,563,721,987,653,461,598,746,231,905,607,541,128,975,231.$$

302° *Lightning calculator.* Zacharias Dase was born in Hamburg in 1824 and died at the early age of thirty-seven in 1861; he was perhaps the most extraordinary mental calculator who ever lived. Among his performances were the mental calculation of the product of two 8-digit numbers in 54 seconds, of two 20-digit numbers in 6 minutes, of two 40-digit numbers in 40 minutes, and of two 100-digit numbers in 8 hours and 45 minutes. He mentally computed the square root of a 100-digit number in 52 minutes. In 1844 he found π correct to 200 decimal places using Gregory's series in connection with the relation

$$\pi/4 = \arctan \tfrac{1}{2} + \arctan \tfrac{1}{5} + \arctan \tfrac{1}{8}.$$

Dase used his powers more worthily when he constructed a seven-place table of natural logarithms and a factor table of all numbers between 6,000,000 and 9,000,000.

Like all calculating prodigies, Dase had a remarkable memory, and a couple of hours after one of his performances he could repeat all the numbers mentioned in it. He could, at a quick glance, state the number (up to about 30) of the books in a case, sheep in a flock, etc. When some dominoes were placed before him on one occasion, with only a second's look he gave their sum of 117. Once, when asked how many letters there were in a certain line of print, he instantly gave the correct number, 63.

Dase's unusual brilliance did not go beyond reckoning with numbers. Although he received a fair education and was afforded opportunity to develop his extraordinary powers, he made little progress. Throughout his life he remained ignorant of geometry and of any language other than German. Most observers found him dull.

AUGUSTUS DE MORGAN

ALTHOUGH Augustus De Morgan (1806–1871) was born in India (his father having been connected with the East India Company), he was brought up and educated in England, graduating as fourth wrangler from Trinity College. His refusal to submit to the required religious tests barred him from holding a fellowship at either Cambridge or Oxford. As a result he was appointed professor of mathematics at the age of twenty-two at the newly established London University. There he remained, except for short periods of resignation brought on by instances of infringement of academic freedom. He was a conscientious champion of religious and intellectual toleration; he was also a teacher and a writer of unusual ability. As a lover of conundrums and of mathematical and scientific curiosa, he assembled his delightful *Budget of Paradoxes*, a satire on "pseudomaths" that was posthumously edited and published by his widow. It is generally believed that if De Morgan had not indulged in so many scattered interests and had concentrated on fewer fields, he would have contributed more deeply to mathematics, but he surely would not have been nearly so interesting a person. His many witticisms, insights, and remarks have made him one of the most quoted of all mathematicians.

97

303° *De Morgan's useless eye.* One of De Morgan's eyes was sightless from birth, and because of this he suffered a number of cruel practical jokes by some of his schoolmates. This is perhaps the chief reason why as a boy he did not join in sports. Later, taking no part in college athletics either, he adopted the flute as his recreation and learned to play the instrument exquisitely.

It has been held by some psychologists that the perception of distance and of dimensionality depends upon the action of two eyes. De Morgan testified that in his belief he perceived distance and solidity with his one eye just as well as normal two-visioned people.

304° *Conscientious champion and great teacher.* Oxford and Cambridge Universities were so guarded by theological tests that no Jew or Dissenter from the Church of England could hold a position in those institutions. Accordingly, a body of liberal-minded men decided to meet the difficulty by establishing a university in London based on religious neutrality. De Morgan, who was then only twenty-two years of age, was appointed Professor of Mathematics at the new London University.

The relations, in the new university, of the Council of management, the Senate of professors, and the body of students were ill defined. As a result, a dispute arose between the professor of anatomy and his students, and as a consequence of the action taken by the Council a number of professors, headed by De Morgan, resigned. Another professor of mathematics was appointed, but a few years later he accidentally drowned. On this, De Morgan, whose reputation as a teacher was superb, was invited to return to his former chair.

The same body of reformers who had founded London University also founded about the same time the Society for the Diffusion of Useful Knowledge. The object of the society was to make scientific and other knowledge easily available by means of cheap but clear treatises written by outstanding writers of the time. De Morgan became one of the society's most voluminous and effective writers. The society published De Morgan's excellent *Differential and Integral Calculus*, and also the famous *Penny Cyclopedia*, about a sixth of the articles of which were written by De Morgan. The *Penny Cyclopedia* was issued in penny numbers.

305° *The London Mathematical Society.* De Morgan had a son George, who distinguished himself in mathematics at both University College and the University of London.* George, and a like-minded friend of his, conceived the idea of founding a mathematical society in London for the reading and discussion of mathematical papers. The first meeting was held in University College; Augustus De Morgan was elected the first president and his son George De Morgan was elected the first secretary. This was the beginning of the London Mathematical Society.

306° *De Morgan and the actuary.* De Morgan was once explaining to an actuary the chance that a certain proportion of a group of people would still be alive at the end of a given period of time. In his explanation he quoted the actuarial formula, which involves π. Upon being asked what π is, De Morgan said that it stood for the ratio of the circumference of a circle to its diameter. At this his acquaintance, who had so far listened attentively to the explanation, interrupted and exclaimed: "My dear friend, that must be a delusion. What can a circle have to do with the number of people alive at a given time?"

307° *De Morgan and the solidus notation.* When writing on the calculus of functions for the *Encyclopedia Metropolitana* (1845), De Morgan proposed the use of the slant line or "solidus" for printing fractions. This proposal was adopted by G. G. Stokes in 1880 and Cayley wrote Stokes, "I think the 'solidus' looks very well indeed...; it would give you a strong claim to be President of the Society for the Prevention of Cruelty to Printers."

Actually, essentially the solidus notation appeared earlier in Spanish America. In the *Gazetas de Mexico* (1784), page 1, Manuel Antonio Valdes used a curved line resembling the modern sign of integration, thus: $3\int 4$. Henri Cambuston in 1843, at Monterey,

---

* The University of London, which is not to be confused with London University, at which De Morgan received his initial appointment, was founded by the government about ten years later for the purpose of granting degrees after examination without any requirements as to residence. London University became affiliated as a teaching college with the University of London, and changed its name to University College.

California, brought out a small text in arithmetic in which he employed a similar curved line for writing fractions.

308° *Ten quotes from De Morgan.* The following quotes illustrate both De Morgan's insight and his wit.

1. The moving power of mathematical invention is not reasoning but imagination.

2. Remember this, the rule for giving an extempore lecture is— let the mind rest from the subject entirely for an interval preceding the lecture, after the notes are prepared; the thoughts will ferment without your knowing it, and enter into new combinations; but if you keep the mind active upon the subject up to the moment, the subject will not ferment but stupefy.

3. It is easier to square the circle than to get round a mathematician.

4. We know that mathematicians care no more for logic than logicians for mathematics. The two eyes of exact science are mathematics and logic: the mathematical sect puts out the logical eye, the logical sect puts out the mathematical eye; each believing that it can see better with one eye than with two. [Today this lack of recognition has been overcome.]

5. German intellect is an excellent thing, but when a German product is presented it must be analysed. Most probably it is a combination of intellect (I) and tobacco-smoke (T). Certainly I_3T_1 and I_2T_1 occur; but I_1T_3 is more common, and I_2T_{15} and I_1T_{20} occur. In many cases metaphysics (M) occurs and I hold that $I_aT_bM_c$ never occurs without $b + c > 2a$.

Be careful, in analysing the compounds of the three, not to confound T and M, which are strongly suspected to be isomorphic. Thus, $I_1T_3M_3$ may easily be confounded with I_1T_6. As far as I dare say anything, those who have placed *Hegel*, *Fichte*, etc., in the rank of the extenders of *Kant* have imagined T and M to be identical.

6. Geometry is the application of strict logic to those properties of space and figure which are self-evident, and which therefore cannot be disputed. But the rigor of this science is carried one step further; for no property, however evident it may be, is allowed to pass without demonstration, if that can be given. The question is therefore to demon-

strate all geometrical truths with the smallest possible number of assumptions.

7. Common integration is only the *memory of differentiation*...the different artifices by which integration is effected are changes, not from the known to the unknown, but from forms in which memory will not serve us to those in which it will.

8. Great fleas have little fleas upon their backs to bite 'em,
And little fleas have lesser fleas, and so *ad infinitum*.
And the great fleas themselves, in turn, have greater fleas to go on;
While these again have greater still, and greater still, and so on.

9. The pseudomath is a person who handles mathematics as a monkey handles the razor. The creature tried to shave himself as he had seen his master do; but, not having any notion of the angle at which the razor was to be held, he cut his own throat. He never tried it a second time, poor animal! but the pseudomath keeps on in his work, proclaims himself clean shaved, and all the rest of the world hairy.

The graphomath is a person who, having no mathematics, attempts to describe a mathematician. Novelists perform in this way: even Walter Scott now and then burns his fingers. His dreaming calculator, Davy Ramsay, swears "by the bones of the immortal Napier." Scott thought that the philomaths worshipped relics: so they do in one sense.

10. Proof requires a person who can give and a person who can receive....

> A blind man said, As to the Sun,
> I'll take my Bible oath there's none;
> For if there had been one to show
> They would have shown it long ago.
> How came he such a goose to be?
> Did he not know he couldn't see?
> Not he.

ALBERT EINSTEIN

ALBERT Einstein was born in Ulm, Germany, in 1879. In 1933, due to events in Nazi Germany, he accepted life membership in the Institute for Advanced Study at Princeton, New Jersey, and in 1940 became a citizen of the United States. He died at Princeton in 1955. His work, all done in the field of mathematical physics, illustrates the close relationship between abstract mathematics and scientific theory that has characterized the twentieth century.

Einstein was a shy and unassuming man—full of compassion, integrity, and humor—and a great body of anecdotes and legends have grown up about him. Most of these stories try to show contrasts in his personality and make-up—to show that he was both simple and deep, both naive and complex.

309° *Einstein and his hat.*   Einstein was noted for his somewhat careless and unconventional dress habits. It has been said, for example, that on some rainy days he went out in public with his overshoes merely drawn over his bare feet.

One threatening morning as Einstein was about to leave his house in Princeton, Mrs. Einstein advised him to take along a hat. Einstein, who rarely used a hat, refused.

"But it might rain!" cautioned Mrs. Einstein.

"So?" replied the mathematician. "My hair will dry faster than my hat."

310° *Einstein's early public address in America.*   Here is a story to show how naive Einstein was on occasion. Shortly after he moved to America, he was inveigled into giving an address before a group of mathematicians at Princeton University. It took some coaxing, for, with characteristic twisting and squirming, he claimed he had nothing to say that the audience wouldn't already know. At last he agreed to talk on some aspects of tensor analysis, a tool essential to the mathematical treatment of relativity theory. A small card, announcing the speaker, time, and place, was put up on the notices board of Fine Hall, where the talk was to be given.

When the day for the address arrived, Princeton University campus was filled with automobiles, suggesting a Princeton–Yale football game, and great crowds of people were milling about Fine Hall, trying to get into the small auditorium there. It turned out that the little card posted in Fine Hall, and intended only for the interested mathematicians, was read by some students. These informed other students. Students wrote home to parents, and the parents came, picking up friends on the way. The townspeople of Princeton also arrived. Everyone wanted to hear the great man speak.

Einstein was led through the shoving crowd and placed in a seat in the front row of the little auditorium, to await introduction at the proper moment. Swiveling his head and looking about in surprise at the excited and pushing crowd struggling to get into the hall he exclaimed: "I never realized that in America there was so much interest in tensor analysis."

311° *Einstein and his blind friend.* This story shows how complex Einstein could be. Not long after his arrival in Princeton he was invited, by the wife of one of the professors of mathematics at Princeton, to be guest of honor at a tea. Reluctantly, Einstein consented. After the tea had progressed for a time, the excited hostess, thrilled to have such an eminent guest of honor, fluttered out into the center of activity and with raised arms silenced the group. Bubbling out some words expressing her thrill and pleasure, she turned to Einstein and said: "I wonder, Dr. Einstein, if you would be so kind as to explain to my guests in a few words, just what is relativity theory?"

Without any hesitation Einstein rose to his feet and told a story. He said he was reminded of a walk he one day had with his blind friend. The day was hot and he turned to the blind friend and said, "I wish I had a glass of milk."

"Glass," replied the blind friend, "I know what that is. But what do you mean by milk?"

"Why, milk is a white fluid," explained Einstein.

"Now fluid, I know what that is," said the blind man. "but what is white?"

"Oh, white is the color of a swan's feathers."

"Feathers, now I know what they are, but what is a swan?"

"A swan is a bird with a crooked neck."

"Neck, I know what that is, but what do you mean by crooked?"

At this point Einstein said he lost his patience. He seized his blind friend's arm and pulled it straight. "There, now your arm is straight," he said. Then he bent the blind friend's arm at the elbow. "Now it is crooked."

"Ah," said the blind friend. "Now I know what milk is."
And Einstein, at the tea, sat down.

312° *An Einstein legend.* An apple falling from a tree and hitting Isaac Newton on the head is said to have suggested to Newton the idea of gravitational force. Perhaps on a par with this story is the following, which purports to tell what led Einstein to the theory of the gravitational field. One day, seeing a workman fall from a building and land unhurt on a pile of straw, Einstein asked the workman if on the way down he had noticed the tug of the "force" of gravity. On being informed that no force had tugged, Einstein immediately saw that "gravitation" can be replaced by an acceleration of the observer's reference system.

313° *Some Einstein quotes.*   1. "Common sense," Einstein once remarked, "is nothing more than a deposit of prejudice laid down in the mind before you reach eighteen."

2. Religious thought is an attempt, he said, "to find out where there is no door."

3. In reply to critics who preferred the probabilistic interpretation of quantum theory as the proper basis of an understanding of physics, he said, "I cannot believe God plays dice with the universe."

4. He once summed up his general outlook on the world by saying, "God is subtle, but he is not malicious."

5. He maintained that in science, though the world can be understood in terms of reason, the criteria for the acceptance of a theory are, in the last analysis, aesthetical.

6. In regard to the real nature of scientific truth in contrast to mathematical truth, Einstein said, "As far as the laws of mathematics refer to reality, they are not certain; and as far as they are certain, they do not refer to reality."

SKIPPING THROUGH THE F'S

314° *A lethal factor table.* Extensive factor tables are valuable in researches on prime numbers. Such a table for all numbers up to 24,000 was published by J. H. Rahn in 1659, as an appendix to a book on algebra. In 1668, John Pell of England extended this table up to 100,000. As a result of appeals by the Swiss-German mathematician J. H. Lambert, an extensive and ill-fated table was computed by a Viennese schoolmaster named Felkel. The first volume of Felkel's computations, giving factors of numbers up to 408,000, was published in 1776 at the expense of the Austrian imperial treasury. But there were very few subscribers to the volume, and so the treasury recalled almost the entire edition and converted the paper into cartridges to be used in a war for killing Turks!

In the nineteenth century, the combined efforts of Chernac, Burckhardt, Crelle, Glaisher, and the lightning calculator Dase led to a table covering all numbers up to 10,000,000 and published in ten volumes. The greatest achievement of this sort, however, is the table calculated by J. P. Kulik (1773–1863), of the University of Prague. His as yet unpublished manuscript is the result of a twenty-year hobby, and covers all numbers up to 100,000,000. The best available factor table is that of the American mathematician D. N. Lehmer (1867–1938); it is a cleverly prepared one-volume table covering numbers up to 10,000,000. Lehmer has pointed out that Kulik's table contains errors.

315° *What became of Karl Feuerbach?* Geometers universally regard the so-called *Feuerbach theorem* as undoubtedly one of the most beautiful theorems in the modern geometry of the triangle. This theorem concerns itself with five important circles related to a triangle. These five circles are the *incircle* (or circle inscribed in the triangle), the three *excircles* (or circles touching one side of the triangle and the other two produced), and the *nine-point circle* (or circle passing through the three midpoints of the sides of the triangle).* Now the Feuerbach theorem says that *for any triangle, the nine-point circle is tangent to the incircle and to each of the three excircles of the triangle.*

A figure (see Figure 37) illustrating this theorem is very attractive, and, if drawn large and then framed, makes a fine decoration on the wall of a high school geometry classroom.

---

* The nine-point circle of a triangle also passes through the three feet of the altitudes of the triangle and the three points bisecting the joins of the orthocenter (the point where the altitudes of the triangle concur) to the three vertices of the triangle.

The theorem was first stated and proved by Karl Wilhelm Feuerbach (1800–1834) in a little work of his published in 1822. It constitutes
his only claim to fame in the field of mathematics. Why did he not
produce further? What became of him? Why did he die at so young
an age as thirty-four? The answers to these questions constitute quite a
tale.

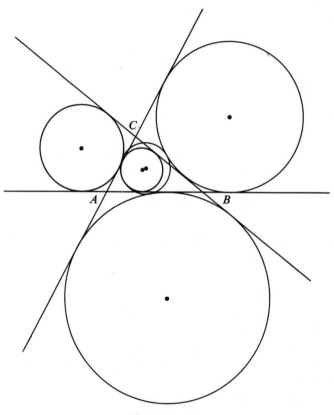

FIGURE   37

Karl, the third son in a family of eleven children, was born in Jena
on May 30, 1800. His father was a famous German jurist, becoming in
1819 the president of the court of appeals in Ansbach. Karl studied at
both the University of Erlangen and the University of Freiburg, and
in 1822 published his little book containing the beautiful theorem. He

received an appointment as professor of mathematics at the Gymnasium at Erlangen, and his father and family were very proud of him.

Then, very suddenly and without warning, one day while walking to school, Karl was arrested. He and nineteen other young men were rounded up and imprisoned in the New Tower in Munich, and held there incommunicado for several months. They were arrested presumably because of the political nature of the activities of an organization to which they had belonged as undergraduates.

During the incarceration, Karl became obsessed with the idea that only his death could free his companions. He accordingly one day slashed the veins in his feet, but before he bled to death he was discovered and removed in an unconscious state to a hospital. There, one day, he managed to bolt down a corridor and leap out of a window. But he fell into a deep snowbank and thus failed to take his life, though he did emerge permanently crippled so that later he looked like a walking question mark.

Shortly after his hospital adventure, Karl was paroled in the custody of a former teacher and friend of the family. One of the other nineteen young men died while in prison, and it was not until after fourteen months that a trial was held and the men were vindicated and released. King Maximilian Joseph took great pains to assist the young men in returning to normal life.

Karl was appointed professor of mathematics at the Gymnasium at Hof, but before long he suffered a breakdown and was forced to give up his teaching. By 1828 he recovered sufficiently to resume teaching, this time at the Gymnasium at Erlangen. However, one day he appeared in class with a drawn sword and threatened to behead any student who failed to solve some equations he had written on the blackboard. This wild and unbecoming act earned him permanent retirement. He gradually withdrew from reality, allowed his hair, beard, and nails to grow long, and became reduced to a condition of vacant stare and low unintelligible mumbling. After living in retirement in Erlangen for six years, he quietly died on March 12, 1834.

316° *Enraptured over heat.* Jean-Baptiste-Joseph Fourier was born on March 21, 1768, at Auxerre, France, the son of a tailor. He was orphaned at eight, cared for by a charitable lady, and sent to the local military school run by the Benedictine Order. At the age of twelve he wrote stirring sermons for some of the leading church dignitaries of Paris, and thought of becoming a priest. Instead he became a teacher of mathematics, first at his local school and then later at the École Normale and the École Polytechnique. In 1798 he enthusiastically joined Gaspard Monge in Napoleon's ill-fated Egyptian campaign.

Of his scholarly achievements, Fourier is today best known for his celebrated *Théorie analytique de la chaleur* (Analytical Theory of Heat) of 1822, which proved to be a landmark in mathematical physics. This work was an extension of ideas that ten years earlier won him the Académie prize for an essay on the mathematical theory of heat, and it was in this work that Fourier contributed the idea that almost any function $y = f(x)$ can be represented by a *trigonometric* (or, as it is called today, a *Fourier*) *series*. Lord Kelvin described the work as "a great mathematical poem."

Fourier's experience in Egypt, and maybe his work on heat, later induced within him a curious habit. He became convinced that desert heat is the ideal condition for good health. He accordingly clothed himself in many layers of garments and lived in rooms of unbearably high temperature. Some believe that this obsession with heat may have hastened his death, the more immediate cause of which was heart disease. He died, thoroughly cooked, on May 16, 1830, in his sixty-third year.

Perhaps Fourier's most quoted sentence (it appeared in his early work on the mathematical theory of heat) is: "The deep study of nature is the most fruitful source of mathematical discoveries."

317° *Benjamin Franklin and mathematics.* [The following is adapted, with permission, from an article by Richard W. Feldmann, Jr., of the same title that appeared in the Historically Speaking section of *The Mathematics Teacher*, February, 1959, pp. 125–127.]

Although Benjamin Franklin (1706–1790) was one of early America's foremost inventors and natural philosophers, he cannot be classified as a great mathematician. His belief that the only value of science is its practical side is shown by his statement, "Nor is it of much importance to us to know the manner in which nature executes her laws. 'Tis enough if we know the laws themselves. 'Tis of real use to know that china left in the air unsupported will fall and break, but how it comes to fall, and why it breaks, are matters of speculation. 'Tis a pleasure indeed to know them, but we can preserve our china without it."

When nine years old, according to his *Autobiography*, Franklin was enrolled in Mr. George Brownell's school for writing and arithmetic, where he "failed in the arithmetic, and made no progress in it." While he was serving his apprenticeship in his brother's printing shop, Franklin read Cocker's book on arithmetic and two works on navigational geometry "with great ease."

The only published article on mathematics by Dr. Franklin appeared in the October 30, 1735, issue of his *Pennsylvania Gazette*. In "On the usefulness of the mathematics," the practical aspect is stressed. Arithmetic is said to be for merchants, shopkeepers, and traders; geometry is for architects, astronomers, geographers, mariners, and surveyors. The article concludes with a sketch of mathematics in ancient history.

In an attempt to establish an educational academy in Philadelphia in 1749, Dr. Franklin published his *Proposals Relating to the Education of Youth in Pennsylvania*, in which he suggests a curriculum for the academy. The only mathematics included are "arithmetick, accounts, and some of the first principles of geometry and astronomy." Accounting is stressed in a footnote with the comment that the keeping of accounts is essential for all gentlemen.

The remainder of his mathematical efforts are found in two letters written to Peter Collinson of England about 1750. The first of these concerns magic squares, that is, square arrays of numbers such that the sums of the numbers in each row, in each column, and in each of the two principal diagonals are identical. In an n by n square, the elements are the integers from 1 through n^2. In Franklin's letter there are two magic squares, an 8 by 8 and a 16 by 16. Unfortunately, according to Albert Chandler in the *Journal of the Franklin Institute* [251 (1951), 415–422], the 16 by 16 square was set up wrongly by a printer in one of the early editions of Franklin's *Works*, and this incorrect square turns up in the later editions. Mr. Chandler has provided a corrected square, which is reproduced in Figure 38. Dr. Franklin referred to this one as the "most magically magical of any magic square ever made by any magician," because it has additional properties. The bent diagonals (indicated in Figure 38) also total the same, namely 2056, as the rows, columns, and principal diagonals. Bent diagonals also run from top to bottom, left to right, and right to left, but, for simplicity, are not shown. The elements in any 4 by 4 subsquare also add up to 2056. Although proud of his squares, he speaks of the time consumed in constructing them as time "which I still think I might have employed more usefully."

200	217	232	249	120	105	88	73	56	41	24	9	136	153	168	185
58	39	26	7	138	151	170	183	202	215	234	247	122	103	90	71
198	219	230	251	118	107	86	75	54	43	22	11	134	155	166	187
60	37	28	5	140	149	172	181	204	213	236	245	124	101	92	69
201	216	233	248	121	104	89	72	57	40	25	8	137	152	169	184
55	42	23	10	135	154	167	186	199	218	231	250	119	106	87	74
203	214	235	246	123	102	91	70	59	38	27	6	139	150	171	182
53	44	21	12	133	156	165	188	197	220	229	252	117	108	85	76
205	212	237	244	125	100	93	68	61	36	29	4	141	148	173	180
51	46	19	14	131	158	163	190	195	222	227	254	115	110	83	78
207	210	239	242	127	98	95	66	63	34	31	2	143	146	175	178
49	48	17	16	129	160	161	192	193	224	225	256	113	112	81	80
196	221	228	253	116	109	84	77	52	45	20	13	132	157	164	189
62	35	30	3	142	147	174	179	206	211	238	243	126	99	94	67
194	223	226	255	114	111	82	79	50	47	18	15	130	159	162	191
64	33	32	1	144	145	176	177	208	209	240	241	128	97	96	65

FIGURE 38

In the second letter, Dr. Franklin produced a magic circle, which is shown in the frontispiece of this volume. Its magical properties are that: (1) the numbers in any radial row, added to the central 12, total 360, the number of degrees in a circle, (2) the numbers in the centrally concentric circles, added to the central 12, total 360, (3) the numbers in the centrally concentric circles and lying above (or below) the horizontal diameter, added to half the central 12, total 180, the number of degrees in half a circle, (4) the numbers in any of the twenty eccentric circles centered at A, B, C, or D, added to the central 12, total 360, (5) any four adjoining numbers in two adjacent centrally concentric circles and two adjacent sectors, added to half the central 12, total 180.

An indication that Benjamin Franklin's mathematical abilities were intuitive is his statement that the population of the United States will double itself every twenty-five years. He seems to have made this assumption with almost no data as a guide. As an example, the 1790 census recorded 3,929,214 people. By calculation the 1890 census should have been 62,867,424. By actual count, it proved to be 62,947,714, an error of only 0.13 percent.

318° *The "Little Giant."* Chirin Guido Fubini (1897–1943), the head of the great Italian school of projective differential geometry, spent his later years in Princeton at the Institute for Advanced Study. There he became nicknamed the "Little Giant," because of his small body but large mind.

It was known that Fubini had three great loves: his family, his mathematics, and his teaching. His son Gino has told a story showing that Fubini displayed these three loves even in the last minutes of life. On the morning of June 6, 1943, when Fubini was quite ill, Gino asked him to solve a mathematical problem. The "Little Giant" immediately immersed himself in the involved equations and calculations, temporarily forgetting his ailments, and very quickly and with great clarity solved the problem for Gino. A few moments later he died.

CARL FRIEDRICH GAUSS

CARL Friedrich Gauss was the greatest mathematician of the nineteenth century, and is usually ranked, with Archimedes and Isaac Newton, as one of the three greatest mathematicians of all time. He was born in Brunswick in 1777. His father was a hardworking laborer with stubborn views and little appreciation of the appropriate schooling for the boy, but his mother, though herself uneducated, did all she could to encourage her son in his studies. At fifteen, Gauss matriculated at the college in Brunswick, through the financial assistance of the Duke of Brunswick, and three years later, still with the Duke's help, he entered

Göttingen University. Gauss remained at Göttingen the rest of his long life, living at the observatory there. He traveled little and never left Germany. Though a mathematician of incredibly fertile mind, he published relatively little—but what he did publish was of the highest quality and was written with enviable finish and mathematical elegance. He died at Göttingen in 1855.

319° *Gauss's precocity.* Gauss very early in life exhibited a remarkable cleverness with numbers, becoming a "wonder child" at the age of two. There are a couple of oft-told stories illustrating the boy's unusual ability.

One of the stories tells how on a Saturday evening Gauss's father was making out the weekly payroll for the laborers of the small brick-laying business that he operated in the summer. The father was quite unaware that his young three-year-old son Carl was following the calculations with critical attention, and so was surprised at the end of the computation to hear the little boy announce that the reckoning was wrong and that it should be so and so instead. A check of the figures showed that the boy was correct, and on subsequent Saturday evenings the youngster was propped up on a high stool so that he could assist with the accounts. Gauss enjoyed telling this story later in life, and used to joke that he could figure before he could talk.

The other story dates from Gauss's schooldays, when he was about ten years old. At the first meeting of the arithmetic class, Master Büttner asked the pupils to write down the numbers from 1 through 100 and add them. It was the custom that the pupils lay their slates, with their answers thereon, on the master's desk upon completion of the problem. Master Büttner had scarcely finished stating the exercise when young Gauss flung his slate on the desk. The other pupils toiled on for the rest of the hour while Carl sat with folded hands under the scornful and sarcastic gaze of the master. At the conclusion of the period, Master Büttner looked over the slates and discovered that Carl alone had the correct answer, and upon inquiry Carl was able to explain how he had arrived at his result. He said, "$100 + 1 = 101$, $99 + 2 = 101$, $98 + 3 = 101$, etc., and so we have as many 'pairs' as there are in 100. Thus the answer is 50×101, or 5050."

320° *The near loss of a genius.* Very likely there were accidents that might have cut off the lives of Archimedes and Newton long before these men attained their great renown. Such an accident occurred to Gauss in his earliest childhood. The canal running by his home became filled to overflowing by a spring freshet. Playing near the water's edge, the little boy fell in. He was saved from drowning by a workman who happened to pass by at that moment.

321° *How Gauss was won to mathematics.* Though Gauss showed a great predilection for mathematics at an early age, it was not at all certain that he would make the study of mathematics his life's work, for he also was a gifted linguist and seriously considered the study of languages instead. But on March 30, 1796, an event occurred that definitely decided Gauss in favor of mathematics. On that day, just a month before his twentieth year, Gauss made the remarkable discovery that those, and only those, regular polygons having a prime number of sides p can be constructed with straightedge and compasses if and only if p is of the form $2^{2^n} + 1$. Now the ancient Greeks had found how to construct with straightedge and compasses regular polygons of 3, 4, 5, 6, 8, 10, and 15 sides. If in the formula $p = 2^{2^n} + 1$ we set $n = 0$ and 1, we obtain the primes 3 and 5 respectively—cases already known to the Greeks. For $n = 2$, we find $p = 17$, which is a prime number. Therefore Gauss showed that a regular polygon of 17 sides is constructible with straightedge and compasses, which was unknown to the Greeks. Gauss was vastly proud of this discovery, and he said that it induced him to choose mathematics instead of philology as his life work.

There is a report that Gauss, in Archimedean fashion, requested that a regular polygon of 17 sides be inscribed on his tombstone. Although this request was never fulfilled, such a polygon is found on the base of a monument to Gauss erected at his birthplace in Brunswick.

322° *Gauss and his chaffinch.* A small group of friends sat in Gauss's home one day discussing the intelligence of animals. In the group was Georg Heinrich Bode, a young professor of classical languages, who had many wonderful things to say on the subject. He told, in particular, about an accomplished parrot that he had secured in his travels in America and had brought back with him to Germany. This parrot was so wise that Bode had named it Socrates. Gauss listened quietly to the praises of the parrot until Bode claimed that Socrates could even answer questions asked of him in Greek. At this Gauss smilingly remarked that though he had never taught his pet chaffinch Hansi any Greek, he had succeeded in teaching the little bird a few words of his native Brunswick dialect, and that the bird had learned to use these words cleverly. As an example, Gauss said that only a few days ago he had held out to the bird a cigar and a pipe and asked, "Which shall I smoke, Hansi?" After a brief meditation, Hansi replied with *"Piep"* (Low German for "pipe").

323° *Gauss and languages.* Gauss mastered languages with great facility. This pursuit of languages became more than just a hobby; he would acquire a new language to test the plasticity of his mind as he grew older, and he considered the exercise as of value in keeping his mind young. At the age of sixty-two he started an intensive self-study of Russian. Within two years he read the language fluently and spoke it perfectly.

324° *Gauss's scientific diary.* On the same day that Gauss discovered his findings on the constructibility of regular polygons with straightedge and compasses (the discovery that caused him to take up mathematics rather than philology as his life's work), he started his scientific diary, or *Notizenjournal.* This scientific diary, which contains 146 entries, is a valuable document in the history of mathematics, for it shows how early Gauss obtained many deep results that were independently discovered and published by others often many years later.

The first entry of the diary records the discovery about regular polygons.

The entry for July 10, 1796, reads, somewhat cryptically,

$$\text{ΣΥΡΗΚΑ!} \quad \text{num} = \Delta + \Delta + \Delta.$$

Here we have an echo of Archimedes' triumphant "Eureka," and an abbreviated statement that every positive integer is the sum of three triangular numbers (a triangular number is a number of the form

$\frac{1}{2}n(n + 1)$, where n is a nonnegative integer). This is not an easy thing to prove from scratch.

The entry for October 11, 1796, is

Vicimus GEGAN,

and that for April 8, 1799, is

REV.GALEN

These two entries have remained unintelligible enigmas through the years.

All the entries, except the above two, are for the most part clear. The entry for March 19, 1797, shows that Gauss had already at that time discovered the double periodicity of certain elliptic functions (he was not yet twenty years old), and a later entry shows that he had recognized the double periodicity for the general case. This discovery alone, if Gauss had published it, would have earned him mathematical fame. But Gauss never published it!

325° *A discredited story.*   Time and again Gauss refrained from publishing remarkable things he discovered. Why did he do this?

An explanation, which was repeated by W. W. R. Ball in his engaging history of mathematics, says that Gauss submitted a large part of his first great masterpiece, the *Disquisitiones arithmeticae*, as a memoir to the French Academy, only to have it cursorily rejected; the resulting undeserved humiliation cut Gauss so deeply that he resolved never again to attempt publication except of work undeniably above all criticism as to both form and content.

It has now been proved that there is nothing at all to the above defamatory legend. In 1935, officers of the French Academy ascertained, by an exhaustive examination of the records of the Academy, that the *Disquisitiones* had never been submitted to the Academy, let alone rejected by it.

The truth seems to be, as Gauss himself said, that he undertook mathematical and scientific investigations only in response to a deep urging of his nature, and that publication of the results for the instruction of others was a quite secondary matter. And then, too, ideas so stormed his mind, especially just before he was twenty, that he could complete and polish only a fraction of them.

326° *Gauss's seal and motto.*   In his scientific writing, Gauss was a perfectionist. Claiming that a cathedral is not a cathedral until the last piece of scaffolding is removed, he strove to make each of his works complete, concise, polished, and convincing, with every trace of the analysis by which he reached his results removed. He accordingly adopted as his seal a tree bearing only a few fruits and carrying the motto: *Pauca sed matura* (Few, but ripe).

327° *The greatest mathematician in the world.*   Alexander von Humboldt once asked Laplace who was the greatest mathematician in Germany, and Laplace replied, "Pfaff." "But what about Gauss?" asked the astonished von Humboldt. "Gauss," explained Laplace, "is the greatest mathematician in the *world*."

328° *Gauss and the extortionists.*   When the French invaded Germany in 1807, the victors fined the inhabitants more than some of the poor people could afford. Thus Gauss was ordered to make a contribution to the Napoleonic war chest of two thousand francs, a sum quite beyond his ability to pay. When Gauss's astronomer friend Olbers heard of the amount of the fine, he enclosed it in a letter to Gauss along with an expression of indignation that a scholar like Gauss should be subjected to such extortion. Gauss thanked his friend for his generosity and his sympathy, but declined to accept the money and returned it forthwith. Shortly after this, Gauss received a friendly note from the famous French mathematician Laplace, telling that he had paid Gauss's two thousand-franc fine and considered it a great honor to do so. Since Laplace paid the fine in Paris, Gauss was unable to return the money, but nevertheless he declined the assistance. By now word apparently had gotten about that Gauss did not wish to accept charity, for some admirer in Frankfurt anonymously sent Gauss one thousand guilders. Gauss was unable to trace the source of the gift and so was forced to accept it, but from the gift he repaid Laplace the two thousand francs that had been deposited for him in Paris.

329° *Gauss's declaration.* Once, when asked how he was able to accomplish so great a mass of work of such very high quality, Gauss replied, "If others would reflect on mathematical matters as deeply and as continuously as I have, they too would make my discoveries." One recalls that Newton, in reply to a similar question as to how he succeeded in making discoveries surpassing those of his predecessors, said, "By always thinking about them." There seems no doubt that part of the explanation of the astonishing accomplishment of great men lies in their capacity for intense and prolonged concentration on a matter at hand.

330° *The greatest calamity in the history of science.* Gauss had a boundless admiration for Archimedes, and could not understand how Archimedes had failed to invent the positional system of numeration. This oversight on Archimedes' part was regarded by Gauss as the greatest calamity in the history of science and caused him to exclaim, "To what heights would science now be raised if Archimedes had made that discovery!"

331° *Gauss and Sir Walter Scott.* Gauss eagerly read the works of Sir Walter Scott as they came out. In one of the novels Scott made an astronomical error by writing, "the moon rises broad in the northwest." This slip considerably tickled Gauss and he went about for days correcting all the copies of the novel that he could find.

SOME LITTLE MEN

THERE are many instances in the history of mathematics where a lesser man has achieved a touch of fame for some single and relatively minor accomplishment in the field. One thinks, for example, of Karl Feuerbach and his beautiful theorem (Item 315°), the Viennese schoolmaster Felkel and his ill-fated factor table (Item 314°), the Italian Lazzerini and his fantastic calculation of pi by probability (Item 290°), and some others who have already been considered in these pages. In this section we look at a few more of these "little" men.

332° *President Garfield and the Pythagorean Theorem.* A few of our country's presidents have been tenuously connected with mathematics. George Washington was a noted surveyor, Thomas Jefferson did much to encourage the teaching of higher mathematics in the United States, and Abraham Lincoln is credited with learning logic by studying Euclid's *Elements.* More creative was James Abram Garfield (1831–1881), the country's twentieth president, who in his student days developed a keen interest and fair ability in elementary mathematics. It was in 1876, while he was a member of the House of Representatives,

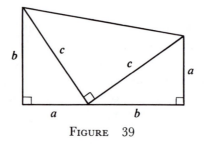

FIGURE   39

and five years before he became President of the United States, that he independently discovered a very pretty proof of the Pythagorean Theorem. He hit upon the proof in a mathematics discussion with some other members of Congress, and the proof was subsequently printed up in the *New England Journal of Education.* Students of high school geometry are always interested to see the proof, which can be presented immediately after the formula for the area of a trapezoid has been covered. The proof depends upon calculating the area of the trapezoid of Figure 39 in two different ways—first by the formula for the area of a trapezoid (as the product of half the sum of the parallel sides times the perpendicular distance between these sides) and then as the sum of three right triangles into which the trapezoid can be dissected. Equating the two expressions so found for the area of the trapezoid, we find (see Figure 39)

$$(a + b)(a + b)/2 = 2[(ab)/2] + c^2/2$$

or

$$a^2 + 2ab + b^2 = 2ab + c^2,$$

whence

$$a^2 + b^2 = c^2.$$

Since a trapezoid, as pictured, exists for any right triangle of legs a and b and hypotenuse c, the Pythagorean Theorem has been established.

333° *Niches in the hall of fame.* The symbol π was used by the early English mathematicians William Oughtred, Isaac Barrow, and David Gregory to designate the circumference, or periphery, of a circle. The first to use the symbol for the ratio of the circumference to the diameter was the English writer William Jones (1675–1749), in a publication in 1706. The symbol was not generally used in this sense, however, until Euler adopted it in 1737.

The symbol $n!$, called *factorial n*, to represent the product

$$(1)(2)(3) \cdots (n - 2)(n - 1)(n),$$

was introduced in 1808 by Christian Kramp (1760–1826) of Strasbourg. He chose the symbol so as to circumvent printing difficulties incurred by a previously used symbol.

Of the considerable mathematical writings and contributions of Jones and Kramp, the above are undoubtedly the only ones by which they will, to any degree, be remembered by posterity.

It is interesting to remark here that Kramp's use of the exclamation point in connection with factorial n, led to the adoption of an associated symbol in combinatorial mathematics. In 1878, W. A. Whitworth introduced *subfactorial n*, defined by

$$n![1 - 1/1! + 1/2! - \cdots + (-1)^n/n!].$$

It represents the number of derangements of a sequence of n objects in which no one of the n objects occupies its original position. Whitworth denoted his subfactorial n by the symbol $\|n$ (since the symbol $\lfloor n$ had been used for factorial n). George Chrystal (1851–1911), of the University of Edinburgh, in Part II of his famous *Textbook of Algebra* (1889), suggested the more convenient notation $n\mathbf{i}$, which has generally been used ever since.

334° *A belated recognition.* An outstanding geometrical problem of the last half of the nineteenth century was to discover a linkage mechanism for drawing a straight line. A solution was finally found in 1864 by a French army officer, A. Peaucellier (1832–1913,) and an announcement of the invention was made by A. Mannheim (1831–1906), a brother officer of engineers and inventor of the so-called Mannheim slide rule, at a meeting of the Paris Philomathic Society in 1867. But the announcement was little heeded until Lipkin, a young student of the celebrated Russian mathematician Chebyshev (1821–1894), independently reinvented the mechanism in 1871. Chebyshev had been trying to demonstrate the impossibility of such a mechanism. Lipkin received a substantial reward from the Russian government, whereupon Peaucellier's merit was finally recognized and he was awarded the great mechanical prize of the Institut de France.

Peaucellier's instrument contains seven bars. In 1874 Harry Hart (1848–1920) discovered a five-bar linkage for drawing straight lines, and no one has been able since to reduce this number of links or to prove that a further reduction is impossible. It has been proved that there exists a linkage mechanism for drawing any given algebraic curve, but that there cannot exist a linkage mechanism for drawing any transcendental curve.

335° *A famous conjecture.* There are many unproved conjectures regarding prime numbers. One of these was made by Christian Goldbach (1690–1764) in 1742 in a letter to the great Swiss mathematician Euler. Goldbach had observed that every even integer, except 2, seemed representable as the sum of two primes. Thus $4 = 2 + 2$, $6 = 3 + 3$, $8 = 5 + 3, \ldots, 48 = 29 + 19, \ldots, 100 = 97 + 3$, and so forth. Though Euler brought his remarkable powers to bear upon this problem, he was unable to resolve the conjecture one way or the other. In fact, to this day the problem remains intractable, though some progress on it has recently been made. In 1931 the Russian mathematician L. Schnirelmann (1905–1938) showed that every positive integer, prime or composite, can be represented as the sum of not more than 300,000 primes! Somewhat later, the Russian mathematician I. M. Vinogradoff (contemporary) showed that there exists a positive integer N such that any integer $n > N$ can be expressed as the sum of at most four primes, but his proof in no way permits us to appraise the size of N.

Goldbach seems to have been an industrious correspondent and to have had the respect of many of the top mathematicians of his day. It was in a letter to Goldbach in 1746, for example, that Euler first announced his remarkable discovery that i^i (where $i = \sqrt{-1}$) is a real number. But today, about the only mention of Goldbach in the history of mathematics concerns his teasing conjecture given above.

336° *An instance of misplaced credit.* In mathematics one often comes across a theorem, a process, a curve, or other mathematical matter, named after the discoverer. In a surprising number of cases, however, the credit has been misplaced, and should really go to an earlier worker in the field. Such is the situation with John Playfair (1748–1819), a Scottish physicist and mathematician of some note in his day. Playfair, in 1795, constructed an edition of Euclid's *Elements* in which Euclid's involved and verbose parallel postulate is replaced by the more succinct and understandable equivalent: "Through a given point can be drawn only one line parallel to a given line." This statement has become known as *Playfair's postulate*, and is the form the parallel postulate takes in most of the modern high school textbooks of geometry. It is almost by this statement alone that John Playfair is known today.

Actually, Playfair's form of the postulate was essentially given earlier—in 1785 by William Ludlam (in his *Rudiments of Mathematics*), and in 1769 by Joseph Fenn (in his Dublin edition of Euclid's *Elements*). Indeed, this form of the postulate was stated as early as the fifth century, by Proclus, in his *Commentary on Euclid, Book I.* There seems to be something pathetic in achieving fame almost solely for the making of a statement already over thirteen hundred years old!

HAMILTON AND HARDY

337° *A patriot, a prodigy, and a lover of animals.* Sir William Rowan Hamilton (1805–1865), by all odds Ireland's greatest bid to fame in the field of mathematics, was an ardent patriot, and he early in life developed a burning ambition to achieve something for the renown of his country. Later in life, after he had gained world fame as a mathematician, he used to say that at the beginning of his century people read French mathematics, but that at the end of it they would be reading Irish mathematics.

As a child, Hamilton was a remarkable prodigy. Under the tutelage of an uncle, the young boy mastered a surprising number of languages. When he was three he was a superior reader of English and was considerably advanced in arithmetic; at four he was a fine geographer; at five he read Latin, Greek, and Hebrew; at eight he added Italian and French, and extemporized fluently in Latin; at ten he commenced a study of oriental languages, first acquiring Arabic and Sanskrit, and then quickly a whole host of other oriental tongues. When he was fourteen, he composed a flowery poem of welcome in Persian for the visiting Persian Ambassador.

But perhaps most to Hamilton's credit was his genuine empathy for animals. He not only loved animals, but, what is regrettably very rare indeed, actually respected animals as man's equal.

338° *A visit to a mathematical shrine.* [The following little adventure, as told by Joseph Ayton of the Fair Lawn Senior High School in New Jersey, is here repeated, with permission, from the Historically Speaking section of *The Mathematics Teacher*, October, 1969.]

During the past summer I enjoyed a vacation in Ireland. Recalling the legend of the circumstances surrounding Sir William Hamilton's invention of noncommutative multiplication for quaternions, I decided to try to find the bridge in Dublin. According to Cajori's *History of Mathematics*, it is the Brougham Bridge (pronounced "Broom") over the Royal Canal, and after some search I located it and photographed it.

As mathematics teachers may know, Hamilton wrestled with the problem of quaternion multiplication for fifteen years, and the story goes that one evening, while walking along the Royal Canal with his wife just before dusk, he was struck by a flash of inspiration. Not having any pencil or paper with him, he took out his pocket knife and scratched the unorthodox multiplication table for quaternions into the stone of the Brougham Bridge. This invention of a noncommutative algebra is regarded by historians as a milestone in the liberation of algebra from its traditional mold; it opened the floodgates of modern abstract algebra.

At the bridge I could not find the original inscription (I wonder whether it is the original bridge anyway; it may have been rebuilt since 1843 because it spans not only the Royal Canal but also some railroad tracks). There is a cement tablet embedded in the stone, however, and it tells the story (see Figure 40).

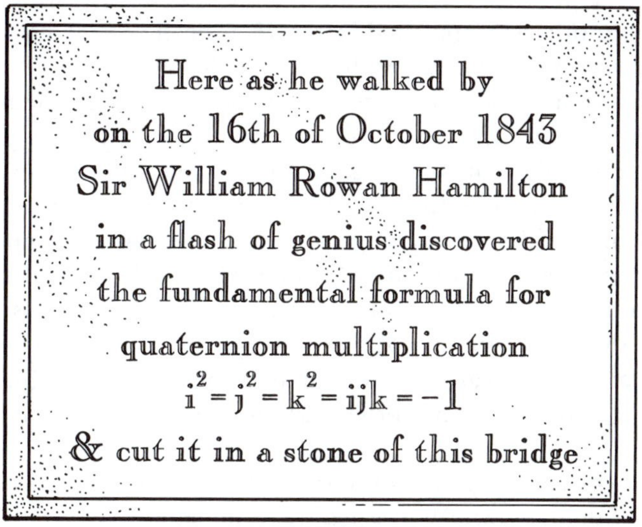

Here as he walked by
on the 16th of October 1843
Sir William Rowan Hamilton
in a flash of genius discovered
the fundamental formula for
quaternion multiplication

$$i^2 = j^2 = k^2 = ijk = -1$$

& cut it in a stone of this bridge

FIGURE 40

The Brougham Bridge was not easy to find. The otherwise very efficient Irish Tourist Information Office in Dublin drew a blank on this one. Their people had never heard of the Brougham Bridge, nor indeed of Sir William Rowan Hamilton either. Apparently it is not considered one of Ireland's greatest tourist attractions. Moreover, the many bridges across the Royal Canal no longer have individual names. A street map of Dublin showed that one of the streets crossing the canal about three miles' bus ride from the center of the city was named Broombridge Street. This proved to be the one.

The ravages of weather and vandals over the years have not been kind to the cement tablet; moreover the Royal Canal itself at this point (ugh!) is not a pretty sight. I have written to the Irish Tourist Office, suggesting that more attention ought to be paid to the memory of one of Ireland's greatest scholars. Great men whose contributions to the race are cultural deserve a place in history at least comparable to that usually accorded to kings and other political figures. All in all, I derived a feeling of great satisfaction from this pilgrimage.

339° *The case of the two Sir William Hamiltons.* [The following is adapted, with permission, from the article, by Howard Eves, of the same title that appeared in the Historically Speaking section of *The Mathematics Teacher*, May, 1963, pp. 348–349.]

There is probably no greater source of confusion in the history of mathematics (indeed, in the history of any subject) than the perpetuation of an error of fact unknowingly made by some authority in the field. Such errors are innocently carried on and multiplied by subsequent writers leaning either on the original authority or on someone who did lean on the original authority. Anyone who works in the history of mathematics soon becomes aware of this situation, for there is hardly a corner of the subject that is not crisscrossed by it. A writer in the history of mathematics just cannot always start from the very beginning; he must constantly borrow from the existing reservoir of written information.

A salient case in point is the frequent confounding of the two Sir William Hamiltons who flourished as contemporaries in the first half of the nineteenth century. The one Sir William Hamilton was born in Glasgow in 1788 and died in Edinburgh in 1856. He was a Scottish baronet and served as a professor of logic and metaphysics at the University of Edinburgh, and he wrote a number of notable papers in the field of his specialty.

The other Sir William Hamilton was born in Dublin in 1805 and died in the same city in 1865. He served as a professor of astronomy at the University of Dublin, published highly original works in mathematics and physics, and was knighted in 1835. This second Sir William Hamilton is more fully known as Sir William Rowan Hamilton, the creator of the calculus of quaternions, and history has recorded him as Ireland's most outstanding mathematician.

It is easy to see how these two Sir William Hamiltons have come to be confounded—they were contemporaries who bore the same name, each was titled, and each was a university professor and a writer of distinguished articles. Much of the confusion attending these two men has been increased by the fact that they had Augustus De Morgan as a common eminent correspondent, and biographers of De Morgan have helped mix matters. The British mathematician, logician, and teacher, Augustus De Morgan, was a brilliant and prolific writer and a widespread correspondent. Being personally interested in both mathematics and logic, he entered into extensive correspondence in connection with these subjects with both the Irish Sir William Hamilton and the Scottish Sir William Hamilton. Biographers of De Morgan have confused the two Sir Williams. This confusion is to be noted, for example, in the article on De Morgan in the eleventh edition of *The Encyclopaedia Britannica*, where the biographer identifies the two Sir William Hamiltons with the Irish one, thus crediting the Irish Sir William with the other Sir William's work in connection with the quantifiers of the predicate in logic. The high and deserved reputation of *The Encyclopaedia Britannica* has caused this error to be repeated.

340° *A trivial relation.* There is a story that G. H. Hardy (1877–1947), one of England's foremost mathematicians of the first half of the twentieth century, once commented in a lecture that a certain mathematical relation was trivial. He then hesitated for a moment and queried, "*Is* it trivial?" Then he excused himself, left the lecture room and went to his office. After twenty minutes of figuring in the office, he returned to the lecture room and announced, "Yes, it is trivial."

Being asked, upon an occasion, if the above story were true, Hardy denied it. He said that the most he would admit to is that he might have said, "This is trivial," hesitated a moment and queried, "*Is* it trivial?" and then, after another pause for thought, said, "Yes, it is trivial." All of which goes to show that it doesn't pay to look too closely into the truth of many an anecdote, if one does not wish to lose the story.

341° *Never criticize the sonatas of archdukes.* There is another story told about Professor G. H. Hardy. Hardy did not much believe in the Ph.D., and he was not above writing a doctoral thesis for a student. On one such occasion the student (who was a foreigner) then asked Hardy to write him a letter saying that this thesis was a good one, so that he might use the letter as a recommendation for a better job. This Hardy declined to do. "However," he said, "take your thesis to Littlewood, and he will write you a letter." The student did so. Littlewood read the thesis, and since it was indeed a good one, he wrote the letter and the student got the job.

342° *The most gigantic gambit conceivable.*   In his beautiful and moving little book, *A Mathematician's Apology*, Hardy makes a penetrating remark about the *reductio ad absurdum*, or indirect, method of reasoning, used so much in mathematics. He says that "It is a far finer gambit than any chess gambit: a chess player may offer the sacrifice of a pawn or even a piece, but a mathematician offers *the game*." *Reductio ad absurdum* emerges as the most stupendous gambit conceivable.

TEN MISCELLANEOUS STORIES

343° *A double anticipation.*   The eighteenth-century Italian geometer and poet, Lorenzo Mascheroni (1750–1800), made the surprising discovery that all compass-and-straightedge constructions, insofar as the given and required elements are points, can be made with the compasses alone, and that the straightedge is thus a redundant tool. Of course, straight lines cannot be drawn with the compasses, but any straight line arrived at in a compass-and-straightedge construction can be determined by the compasses alone by finding two points of the line. This discovery appeared in 1797 in Mascheroni's *Geometria del compasso.* Generally speaking, Mascheroni established his results by using the idea of reflection in a line. In 1890, the Viennese geometer, August Adler (1863–1923), published a new proof of Mascheroni's results, using the inversion transformation.

Then an unexpected event happened. Shortly before 1928, a student of the Danish mathematician Johannes Hjelmslev (1873–1950), while browsing in a bookstore in Copenhagen, came across a copy of an old book, *Euclides Danicus*, published in 1672 by an obscure writer named Georg Mohr. Upon examining the book, Hjelmslev was surprised to find that it contained Mascheroni's discovery, with a different solution, arrived at a hundred and twenty-five years before Mascheroni's publication had appeared.

Inspired by Mascheroni's discovery, the French mathematician Jean-Victor Poncelet (1788–1867) considered constructions with straightedge alone. Now not all compass-and-straightedge constructions can be achieved with only the straightedge, but, curiously enough, in the presence of one circle and its center drawn on the plane of construction, all compass-and-straightedge constructions can be carried out with straightedge alone. This remarkable theorem was conceived by Poncelet in 1822 and then later, in 1833, was fully developed by the Swiss-German geometry genius Jacob Steiner (1796–1867).

It was about 980 that the Arabian mathematician Abû'l-Wefâ (940–998) proposed using the straightedge along with *rusty compasses*, that is with compasses of a fixed opening. In view of the Poncelet–Steiner theorem we need, in fact, use the compasses only once, after which the compasses may be discarded. In 1904, the Italian Francesco Severi went still further, and showed that all we need is an arc, no matter how small, of one circle, and its center, in order thenceforth to accomplish all compass-and-straightedge constructions with straightedge alone.

Then, shortly after the midpoint of the present century, another surprising historical discovery was made. It was shown that the Georg Mohr mentioned above was the author of an anonymously published booklet entitled *Compendium Euclidis Curiosi*, which appeared in 1673 and which in effect shows that all the constructions of Euclid's *Elements* are possible with straightedge and rusty compasses. So the same obscure Danish writer who anticipated the Mascheroni construction theorem also essentially anticipated the Poncelet–Steiner construction theorem.

344° *The Jacobi brothers.* In the 1840's there were two brothers, M. H. Jacobi and C. G. J. Jacobi, and the former completely eclipsed the latter. M. H. achieved fame as the founder of a fashionable quackery called galvanoplastics; C. G. J. was a mere mathematician. During his lifetime, C. G. J. was frequently confused with his better known brother, or, worse, was often congratulated on being the brother of the famous quack. Annoyed by such confusions, C. G. J. would assert, "I am not *his* brother, he is *mine*." And this is where time has ultimately left the relationship, for today M. H. Jacobi is known almost only as the brother of C. G. J. Jacobi, and the latter is regarded as one of the great mathematicians of history, ranking, in Germany during his lifetime, second only to Gauss.

345° *From the lips of C. G. J. Jacobi.* Carl Gustav Joseph Jacobi (1804–1851) was always very generous in his statements about his great contemporaries in the field of mathematics. Of one of Abel's masterpieces he said, "It is above my praise as it is above my own work."

Jacobi was a remarkable teacher, and one of the first to train students in research by lecturing to them on his own latest discoveries. Most students feel that before doing research they should first master what has already been accomplished. To offset this notion, and to stimulate early interest in independent work, Jacobi would deliver the parable: "Your father would never have married, and you would not be born, if he had insisted on knowing *all* the girls in the world before marrying *one*."

Jacobi worked with prodigious energy on his mathematics, so much so that occasionally he endangered his health. To a concerned friend he once replied, "Of course I sometimes endanger my health by overwork, but what of it! Only cabbages have no nerves and no worries, and what do they get out of such perfect well-being?"

Jacobi, in defending pure research against applied research, remarked, "The real end of science is the honor of the human mind."

346° *L. Kronecker and J. P. Morgan.* The eminent German mathematician Leopold Kronecker (1823–1891) and the famous American financier John Pierpont Morgan (1837–1913) were, in a sense, obverses of one another. For Kronecker was a superb businessman who, before he was thirty, amassed a sizable fortune for himself, whereas Morgan showed such outstanding ability in mathematics, when a student in Germany, that his professors tried to induce him to follow mathematics as a career and even offered him a university post in Germany.

347° *Kronecker's toast.* Kronecker was a Pythagorean, as his most often quoted remark testifies: "Die ganzen Zahlen hat Gott gemacht, alles andere ist Menschenwerk."* In this connection, see Item 56°.

---

* "God made the whole numbers, all the rest is the work of man."

348° *Monge's romantic marriage.* The marriage of the French mathematician Gaspard Monge (1746–1818) reflects some eighteenth-century romance. While at a reception, Monge overheard a bounder spitefully slandering a young widow for having rejected him. Though the widow, a Madame Horbon, was unknown to Monge, the gallant mathematician took the slanderer to task and unsuccessfully tried to force a duel. A few months later, at another reception, Monge was captivated by the charm of a young lady, and upon introduction found she was the Madame Horbon he had earlier defended. The two were married in 1777. She survived Monge and did all she could to perpetuate his memory; she was perhaps the only human being who stuck to Monge through all the vicissitudes of his life.

349° *What a released POW brought back.* Jean-Victor Poncelet (1788–1867) served as an officer in the French army and was taken a Russian prisoner of war during Napoleon's retreat from Moscow. During his two years of captivity, with no books at hand, Poncelet planned his great work on projective geometry, which, upon his release and return to France, he published in Paris in 1822. This work constituted the real revival of projective geometry; it gave tremendous impetus to the study of the subject and inaugurated the so-called "great period" in the history of projective geometry.

Poncelet brought back something else with him to France when he was released from his captivity in Russia. The period from about 1100 to 1500 witnessed the battle between the abacists and the algorists, the former supporting the use of the abacus for computational work and the latter advocating the Hindu-Arabic numeral system. By 1500 our present rules of computing won supremacy and in the next hundred years the abacists were almost forgotten. By the eighteenth century no trace of an abacus was to be found in western Europe. Its reappearance, as a curiosity, was due to Poncelet, who brought back a specimen to France after his release.

350° *What Poisson learned by hanging around.* Siméon Denis Poisson (1781–1840) was born at Pithiviers in France, and developed into a top-notch mathematician in the applied fields. As a boy he was put in the care of a nurse. One day, when his father (a private soldier) came to see him, the nurse had gone out and left him suspended by his straps to a nail in the wall—to protect the boy, the nurse said, from the disease and dirt of the floor. Poisson said that his gymnastic efforts when thus suspended caused him to swing back and forth, and it was in this way that he early became familiar with the pendulum, the study of which occupied much of his later life.

351° *Poisson as a keeper of money.*   There is a story that Poisson in 1802 was asked by a young man about to enter into the army to take a bag of money into safe-keeping. Since Poisson was at the moment busily engaged, he told the young man to put the money bag on a shelf. The recruit did so, and to hide the bag Poisson placed a copy of Horace over it. Twenty years later the soldier returned and asked for his money, but Poisson had no recollection of it. "You say you put the money in my hands?" queried Poisson. "No," replied the soldier, "I put it on this shelf and you placed this book over it." The soldier removed the dust-laden copy of Horace and found the bag of money where it had been put twenty years before.

352° *Why there is no Nobel Prize in mathematics.*   There is a Nobel prize in several of the great fields of study, but none in mathematics. The reason for this is interesting. At one time the great Swedish mathematician G. M. Mittag-Leffler (1846–1927) was a man of considerable wealth, and in accumulating his fortune he antagonized a number of people, in particular Alfred Nobel, who founded the five great prizes for annual award for the best work in Physics, Chemistry, Physiology or Medicine, for Idealistic Literary Work, and for the Cause of Universal Peace. At the time the prizes were set up, mathematics was also under consideration. Nobel asked his advisers, if there should be a prize in mathematics, in their opinion might Mittag-Leffler ever win it? Since Mittag-Leffler was such an able and famous mathematician, they had to admit that such would indeed be a possibility. "Let there be no Nobel Prize in Mathematics, then," Alfred Nobel ordered.

J. J. SYLVESTER AND NORBERT WIENER

We conclude this first trip around the Mathematical Circle by telling a few stories about J. J. Sylvester (1814–1897) and Norbert Wiener

(1894–1964). These two men are among the most colorful in the whole history of mathematics, and a fair-sized pamphlet could be devoted to stories about just them alone. As the years have rolled by, some of these stories have become embroidered—but in spirit they still tell much about the characters of the two men.

Since we have ample material for at least two or three more trips around the Mathematical Circle, perhaps at some future and more leisurely time we may set forth further collections of these mathematical stories and anecdotes.

353° *Sylvester's memory.* Cayley and Sylvester, mathematical friends for a long period of years, were in almost every sense antitheses of one another. One of the many ways in which they were opposites was in the matter of memory. Cayley seemed never to forget anything he had once seen or read, whereas Sylvester had difficulty in remembering even his own mathematical inventions. On one occasion Sylvester objected to a mathematical statement made by a companion, insisting that the statement had never been heard of and, moreover, simply could not be true. The companion responded by showing the amazed Sylvester a paper written by Sylvester himself, in which Sylvester had announced his discovery of the concerned statement and had written out its proof.

354° *Sylvester at the University of Virginia.* In 1841 Sylvester accepted an appointment as Professor of Mathematics at the University of Virginia, in America. He entered into his new duties with all the enthusiasm and energy of his youthful twenty-seven years, little suspecting that the whole venture would end in an unpleasant experience only three to four months later.

It soon became apparent that there were some students who resented the presence of a foreigner and a Jew on the faculty, and Professor Sylvester began to suffer annoyances in the classroom. Finally, after three months of growing harassment, Professor Sylvester reported a case of serious disrespect accorded him in the classroom by a Mr. W. H. Ballard. Mr. Ballard was summoned to tell, in the absence of Professor Sylvester, his side of the altercation. His report was quite at variance with that of Professor Sylvester, and he had his view of the affair backed up by his crony Mr. W. F. Weeks. Upon hearing of the introduction of this biased witness, Professor Sylvester protested, and wondered why some more neutral student witness had not been employed instead. In an attempt to carry out Professor Sylvester's wishes such witnesses were next called, but they tended to support Mr. Ballard's report. At this point, the faculty records become obscure, but it seems that finding himself in discord with both the students and his fellow faculty members—hurt and dissatisfied with the lack of a firm faculty decision —Professor Sylvester submitted his unconditional resignation. The resignation was accepted and Sylvester left Virginia. He went to New York City, where he had a brother living, and tried unsuccessfully for about a year to find other gainful employment in America. Failing, he returned, penniless, to England.

355° *Sylvester at Johns Hopkins University.* One would think that Sylvester, having suffered such an unhappy and disastrous visit to America in 1841, had had enough of this country. But in 1876 he once again crossed the Atlantic, this time to take up a position at the newly founded Johns Hopkins University in Baltimore. Sylvester remained at Johns Hopkins for seven years, and those seven years proved to be among the happiest and the most productive of his life. It was during his stay at Johns Hopkins that Sylvester founded, in 1878, the *American Journal of Mathematics.*

There are many delightful and colorful tales told of Sylvester during his tenure at Johns Hopkins. The following, told in the words of Dr. Fabian Franklin (Sylvester's successor in the chair of mathematics at Johns Hopkins, and an eyewitness to the narrative told), illustrates Sylvester's enthusiastic interest in poetry and the laws of verse.

"He [Sylvester] made some excellent translations from Horace and from German poets, besides writing a number of pieces of original verse. The tours de force in the way of rhyming, which he performed while in Baltimore, were designed to illustrate the theories of versification of which he gives illustrations in his book called 'The Laws of Verse.' The reading of the Rosalind poem at the Peabody Institute was the occasion of an amusing exhibition of absence of mind. The poem consisted of no less than four hundred lines, all rhyming with the name Rosalind (the long and short sound of the *i* both being allowed).

The audience quite filled the hall, and expected to find much interest or amusement in listening to this unique experiment in verse. But Professor Sylvester had found it necessary to write a large number of explanatory footnotes, and he announced that in order not to interrupt the poem he would read the footnotes in a body first. Nearly every footnote suggested some additional extempore remark, and the reader was so interested in each one that he was not in the least aware of the flight of time, or of the amusement of the audience. When he had dispatched the last of the notes, he looked up at the clock, and was horrified to find that he had kept the audience an hour and a half before beginning to read the poem they had come to hear. The astonishment on his face was answered by a burst of good-humored laughter from the audience; and then, after begging all his hearers to feel at perfect liberty to leave if they had engagements, he read the Rosalind poem."

356° *Sylvester on music and mathematics.* Sylvester had a keen sense of the kinship of mathematics to the finer arts, and he frequently expressed the relationship in his writing. Thus, in a paper of his, "On Newton's rule for the discovery of imaginary roots," he exclaims in a footnote:

"May not Music be described as the Mathematic of sense, Mathematic as Music of the reason? the soul of each the same! Thus the musician *feels* Mathematic, the mathematician *thinks* Music,—Music the dream, Mathematic the working life—each to receive its consummation from the other when the human intelligence, elevated to its perfect type, shall shine forth glorified in some future Mozart-Dirichlet or Beethoven-Gauss—a union already not indistinctly foreshadowed in the genius and labours of Helmholtz!"

357° *Wiener and his car.* Norbert Wiener, for years a noted Professor of Mathematics at the Massachusetts Institute of Technology, appears to have been a model of the eccentric, super-brainy, absent-minded professor. Many stories have evolved to illustrate one or more of these characteristics. Thus there is the story of the time he drove his car to New Haven to attend a mathematics meeting at Yale University. At the conclusion of the meeting, forgetting he had driven down, he took a bus back to Cambridge. The next morning he went out to his garage to get his car, found that it was not there, and notified the police that while he was away at a mathematics meeting in New Haven someone had stolen his car.

358° *Wiener and the student.*   An oft-told tale about Norbert Wiener concerns his meeting a student while crossing the M. I. T. campus. The student accosted him and asked him a question, and Wiener became engaged in a lengthy answer and explanation. At the conclusion, Professor Wiener looked perplexed and confused. Turning to the student, he asked, "Do you recall, when we met, in which direction I was headed?"

359° *Wiener at a colloquium.*   Norbert Wiener was known for sleeping through colloquium talks and then awakening at the conclusion and making pertinent comments. At one such gathering, Professor Saunders MacLane was talking about something in algebra. At the very end of his talk, MacLane, in a loud voice directed toward the sleeping Wiener, said, "Thus we see that the subject has absolutely nothing to do with ERGODIC THEORY." Wiener immediately awoke and commenced to talk about ergodic theory.

360° *Wiener in the classroom.*   One day, in an undergraduate course, Professor Wiener's class asked him how to do a certain problem. He thought for a moment, then wrote the answer on the blackboard. The class was nonplussed, and finally a bold spirit asked, "But, Professor Wiener, is there any other way to do the problem?" The professor thought again for a moment, then brightened up and said, "Why yes, there *is* another way." Then he wrote the same answer on the board.

INDEX

References are to items, not to pages. A number followed by the letter *p* refers to the historical capsule just preceding the item of the given number (thus 275*p* refers to the historical capsule immediately preceding Item 275°).

INDEX